BURLEIGH DODDS SCIENCE: INSTANT INSIGHTS

NUMBER 07

Mastitis in dairy cattle

burleigh dodds
SCIENCE PUBLISHING

Published by Burleigh Dodds Science Publishing Limited
82 High Street, Sawston, Cambridge CB22 3HJ, UK
www.bdspublishing.com

Burleigh Dodds Science Publishing, 1518 Walnut Street, Suite 900, Philadelphia, PA 19102-3406, USA

First published 2021 by Burleigh Dodds Science Publishing Limited

British Library Cataloguing in Publication Data
A catalogue record for this book is available from the British Library

ISBN 978-1-78676-929-9 (Print)
ISBN 978-1-78676-930-5 (ePub)

DOI 10.19103.9781786769305

Typeset by Deanta Global Publishing Services, Dublin, Ireland

Contents

Chapter 1

Aetiology, diagnosis and control of mastitis in dairy herds

P. Moroni, Cornell University, USA and Università degli Studi di Milano, Italy; F. Welcome, Cornell University, USA; and M.F. Addis, Porto Conte Ricerche, Italy

1 Introduction

Mastitis is one of the most economically important diseases in dairy production. Intramammary infections (IMI) continue to be the most important cause of mastitis in dairy cattle, accounting for 38% of the total costs of the common production diseases (Kossaibati and Esslemont, 1997). In the last decade, several groups have estimated the losses associated with clinical mastitis (CM), and the average costs per case (US$) of Gram-positive, Gram-negative and other microorganisms were $133.73, $211.03 and $95.31, respectively (Cha et al., 2013; Gröhn et al., 2004). These costs include treatment, culling, death and decreased milk production. In addition to reduced cow welfare and increased veterinary costs, episodes of mastitis are associated with reduction of milk production (Bar et al., 2007; Schukken et al., 2009), decreased fertility (Hertl et al., 2010; Santos et al., 2004), and increased culling and death risk (Hertl et al., 2011).

http://dx.doi.org/10.19103/AS.2016.0006.21

Table 1 Definition of intramammary infection and mastitis

	Intramammary infection	Mastitis
International Dairy Federation definition	An infection occurring in the secretory tissue and/or the ducts and tubules of the mammary gland	Inflammation of one or more quarters of the mammary gland, almost always caused by infecting microorganisms

Source: Lopez-Benavidez et al. (2012).

Since mastitis is most often due to a bacterial IMI (Djabri et al., 2002), the terms IMI and subclinical mastitis have so far been used interchangeably (Barkema et al., 1997; Deluyker et al., 2005). According to a recent document by the National Mastitis Council (Lopez-Benavides et al., 2012), the terms 'mastitis' and 'intramammary infection' represent different entities (Table 1), and the definitions provided by the International Dairy Federation should always be used when referring to these conditions (Lopez-Benavides et al., 2012).

Clinical mastitis is an inflammatory response to infection causing visibly abnormal milk (e.g. colour changes, fibrin clots and watery appearance). Assigning a severity score to individual clinical cases along with identification of the pathogen involved (culture result) helps veterinarians to assign specific treatment protocols. If clinical signs include only visible changes in the appearance of milk, notable swelling or a painful udder, the case is classified as mild or moderate in severity. If the inflammatory response includes systemic involvement (fever, anorexia, shock), the case is categorized as severe. If the onset is very rapid, as often occurs with severe clinical cases, it is termed an acute case of severe mastitis. More severely affected cows tend to have more serous secretions in the affected quarter. Clinical cases that fall into the severe category account for 10% to 15% of infections. Long-term recurring persistent cases of the disease are termed chronic. These may show few signs of inflammation between repeated occasional flare-ups of the disease where signs are visible and can continue over periods of several months. Chronic cases of mastitis are often associated with irreversible damage of the udder tissues from the repeated clinical occurrences of the illness, and these cows are usually culled.

Subclinical mastitis is generally caused by the presence of an infection without any apparent sign of local inflammation or systemic involvement. Even if episodes of abnormal milk or udder inflammation may appear, these infections are generally asymptomatic and, if the infection persists as measured by SCC or SCS for at least 2 months, are classified as chronic. The majority of these infections persist for entire lactations or the life of the cow. Although subclinical mastitis implies inflammation within the udder but not necessarily infection, different pathogens, contagious and environmental, are associated with it. Subclinical mastitis caused by *Staphylococcus aureus* is particularly important as cows continue to shed microorganism putting the uninfected portion of the herd at risk.

The microorganisms most frequently causing mastitis can be divided into two groups based on their source:

- contagious pathogens
- environmental pathogens

In most countries, the major pathogens for contagious mastitis are *S. aureus*, *Streptococcus agalactiae* and *Mycoplasma* spp. (Ruegg, 2014). The word 'major' reflects the number of isolates from IMI and the significance of their impact on cow health, milk quality and productivity. These organisms are well adapted to survival and growth in the mammary gland. The infected gland is the main source of these organisms in a dairy herd, and transmission from infected to uninfected quarters and cows occurs mainly during the milking period. Although there are a handful of major pathogens, the predominant pathogenic cause of CM varies among countries, management styles and farms. Mastitis control strategies will need to meet both the specific requirements of an individual country or segment of the dairy industry but also adapt to differences in epidemiology.

There is a wide range of environmental pathogens including *Escherichia coli*, *Klebsiella* and environmental streptococci. The epidemiology of mastitis is evolving and environmental mastitis pathogens are now the main cause of mastitis on many modern dairy farms. These pathogens often cause mild cases of CM, but some can become host adapted and behave similar to contagious pathogens. In the Northeast of the United States we are continuing to see more individual farm situations in which a predominantly environmental mastitis pathogen becomes the principal organism isolated on a dairy. While there are still outbreak situations with a single environmental organism that occur over a brief time period, many more farms are experiencing mastitis events where a predominantly environmental organism persists as the dominant organism for a prolonged period of time with high numbers of chronically infected cows.

2 Indicators of mastitis: somatic cell count

The most established and widely recognized method for mastitis monitoring at the cow and herd level consists in measuring the cells that are present in milk by determining its somatic cell count (SCC). The SCC is defined as the number of cells per millilitre of milk (cells/mL) (Dohoo and Leslie, 1991; Ruegg and Pantoja, 2013). Somatic cells are primarily macrophages, leucocytes (white blood cells) and some epithelial cells from the mammary gland. A small number of immune cells are in fact present in milk in normal physiological conditions, with the function of protecting the udder against bacterial infection. These are derived from blood and consist of macrophages, lymphocytes and polymorphonuclear neutrophils (PMN) (Hamann, 2005). Macrophages are the predominant cell type present in milk from uninfected quarters. Lymphocytes are responsible for immune memory. PMNs are the first defence against an invasion of the mammary gland by pathogenic microorganisms, and are the major determinant of the increase in SCC.

The biological foundation of using SCC as an indicator of mastitis relies on the fact that, when a pathogen enters the udder and triggers an immune response, immune cells are recruited from the circulatory system and the PMN number in milk increases rapidly. As a consequence, the SCC is an approximation of the number of immune cells in milk. Once the infection is cleared, the SCC gradually returns to normal. Different authors have clearly shown that environmental or physiological factors such as lactation number, days in milk (DIM), oestrus, and heat stress have only low effects on SCC from uninfected quarters (Laevens et al., 1997). A quarter with SCC above 200 000 cells/mL in mature cows (100 000 cells/mL in 1st lactation cows) is an indication of an inflammatory response, the quarter is likely to be infected, and the milk has changed properties such as reduced shelf life of fluid milk, reduced yield and lower quality of cheese (Barbano et al., 2006).

The SCC can be measured in bulk tank milk (BMSCC), at cow level with composite samples of all four quarters (CSSCC) and at quarter level (QMSCC). BMSCC values are the reference for defining national and international standards for hygienic production of milk. Regulatory standards for comingled milk (BMSCC) may significantly differ depending on the country, ranging from <400 000 cells/mL (such as in the EU, Australia, New Zealand and Canada) (USDA, 2013) to <500 000 cells/mL (Brazil from 2016), and are currently 750 000 cells/mL in the USA. The BMSCC is used to monitoring udder health at the herd level. The optimal BMSCC is not definitively described, but it is generally considered to be <250 000 cells/mL. Some milk buyers offer milk quality premiums based in part on BMSCC to milk producers to encourage lower BMSCC (Ruegg and Pantoja, 2013).

BMSCC can provide reliable indications at the herd level, but measuring CSSCC or QMSCC is necessary for monitoring udder health at the cow level (De Vliegher et al., 2012). This helps to keep subclinical mastitis under control and to obtain more reliable estimates on mastitis prevalence and incidence. The dynamics of SCC values at both herd and cow level from Dairy Herd Improvement (DHI) programmes (Laevens et al., 1997; Ruegg, 2003) are used in herd management to identify cows that need interventions including culture, treatment, segregation or removal from the herd (Cook et al., 2002; Rhoda and Pantoja, 2012). QMSCC values of 200 000 cells/mL are currently believed to possess a level of specificity sufficient to provide the least diagnostic error in detecting an IMI (Bradley and Green, 2005; Dohoo and Leslie, 1991; Schepers et al., 1997; Schukken et al., 2003), but lower values may be more adequate if a higher sensitivity is desired (Bradley and Green, 2005; Dohoo and Leslie, 1991; Ruegg and Pantoja, 2013; Schepers et al., 1997; Schukken et al., 2003). In general, the following applies: a QMSCC of 100 000 cells/mL or lower indicates absence of mastitis, while a QMSCC of 200 000 cells/mL or higher indicates presence of mastitis, and therefore IMI.

Since SCC values do not follow a normal distribution, their skewed behaviour is usually accommodated by obtaining a linear score (LS), defined also as somatic cell score (SCS), with the logarithmic transformation of the SCC ($LS = \log2 (SCC/100)+3$). The LS enables to account for very SCC high values, and dampens the effect that these would have on arithmetic means.

Somatic cells can be enumerated in milk by means of automated cell counting instrumentation, either in the laboratory or at the milking plant. A wide range of devices are available on the market that can meet different throughputs and requirements. In addition, somatic cells can be assessed with cow-side methods such as the California Mastitis Test (CMT). The CMT provides a qualitative indication of the number of somatic cells in milk, and it is based on a four-compartment paddle and on a reagent containing a detergent and a pH indicator. Milk from the four quarters is collected in the paddle and an aliquot of the reagent of similar quantity is mixed with it. If a high number of cells is present, a visible gelation occurs, due to cell lysis and to the consequent release of their DNA (Whyte et al., 2005). A score ranging from N (negative) to 3 (strongly positive) is then given to the reaction, based on the absence or presence of a visible gelation. Due to its qualitative nature, the CMT is highly subjective and dependent on user experience, especially for SCC below 1 000 000 cells/mL, and therefore it has a low sensitivity. It is however highly cost-effective and practical for verifying the status of individual quarters. Interpreting individual quarter data from the CMT allows managers to select from a number of options including sampling for culture, treatment, segregation of milk dry off or culling to best manage individual cows.

3 Indicators of mastitis: non-cell inflammation markers

Adding to the use of immune cells as indicators of mastitis, other molecules released in milk as a result of an inflammatory process can represent useful, reliable and practical markers. Several enzymes, sugars and salts are already known to increase in milk during mastitis (Pyorala, 2003) but the advances in biomarker discovery methods based on proteomic techniques (Abd El-Salam, 2014; Ceciliani et al., 2014; Reinhardt et al., 2013; Smolenski et al., 2014) have more recently enabled the identification of other protein and peptide candidates that can form the basis for novel laboratory and field assays. In addition, advancements in immunological assays, for both the laboratory and the field, have increased sensitivity and specificity of biomarker detection and can represent inexpensive and practical alternatives (Gurjar et al., 2012; Viguier et al., 2009).

Different enzymes have been exploited as mastitis markers. Lactate dehydrogenase (LDH) is a cytoplasmic enzyme that changes in abundance in mastitic milk (Chagunda et al., 2006; Hiss et al., 2007). LDH can now be assessed in-line in composite milk, and is available for monitoring mastitis in commercial herd management systems such as the DeLaval Herd Navigator™. In addition, a portable LDH activity assay was developed (Hiss et al., 2007) by using dry chemistry and a field spectrophotometer, having performances similar to the laboratory test. Several reports indicate N-acetyl-β-D-glucosaminidase (NAGase), an enzyme released from PMNs during phagocytosis and upon cell lysis, as a reliable mastitis marker (Holdaway et al., 1996; Mattila et al., 1986; Nielsen et al., 2005). Alkaline phosphatase (AP) has also been explored for its ability to indicate mastitis, although with limited success (Babaei et al., 2007; Bogin and Ziv, 1973). Other tests based on enzymes measure an esterase produced by milk cells on a dipstick (PortaSCC). Other non-enzymatic molecules involved in inflammation have promising potential for mastitis detection (Ceciliani et al., 2012; Miglio et al., 2013; Wheeler et al., 2012). Among these, the major bovine acute phase proteins, serum amyloid A (SAA) and haptoglobin (Hp), have been proposed both as mastitis biomarkers (Eckersall et al., 2001; Hiss et al., 2007; O'Mahony et al., 2006) and as markers of milk quality (Åkerstedt et al., 2008). Hp and SAA are produced in the liver, and in small amounts also in the udder (Hiss et al., 2004; McDonald et al., 2001). Their function is mainly antibacterial, and it is exerted by Hp through binding free heme (Eaton et al., 1982) and by SAA through opsonization of Gram-negative bacteria (Larson et al., 2005). Both have potential as early biomarkers.

Cathelicidin is a small protein with direct antimicrobial activity and potent proinflammatory and chemotactic functions released by both neutrophils and epithelial cells upon pathogen sensing (Wiesner and Vilcinskas, 2010; Zanetti, 2005, 2004). Indications on release of cathelicidin in mastitic ruminant milk have been provided by several authors (Addis et al., 2013, 2011; Ibeagha-awemu et al., 2010; Murakami et al., 2005; Reinhardt et al., 2013; Smolenski et al., 2007; Zhang et al., 2015), and a study by Smolenski et al. (2011) has shown promising results concerning cathelicidin abundance in mastitic cow milk as an indicator of mastitis.

Mastitis-related biomarkers can also be detected by means of laboratory and field immunoassays, provided that reliable antibodies are available for warranting assay robustness and reproducibility. ELISAs have been developed for Hp detection in milk and serum (Hiss et al., 2007), and a solid-phase-sandwich ELISA is commercial available for SAA (Tridelta Phase™ range SAA kit, Tridelta Development Ltd, Co. Wicklow, Ireland). A milk pan-cathelicidin sandwich ELISA based on two monoclonal antibodies has also recently been reported (Addis et al., 2016a, 2016b).

As a further implementation, biosensors and immuno-biosensors have been developed for detecting protein markers of mastitis as well as other, non-protein, mastitis-associated molecules. Biosensors are analytical devices based on an immobilized biological material (e.g. an enzyme, an antibody) that interacts with the molecule to be detected (e.g. a small molecule, a protein) producing a measurable physical, chemical or electrical signal. Adding to enabling detection in the field, these sensors make it possible to implement marker measurement on-line. The recent, significant increase of robotic milking would represent a powerful way to implement biosensor-based, on-line mastitis detection strategies. Currently, on-line tests are available, based on the SCC, milk colour determination, or electrical conductivity (EC) (Hovinen et al., 2006; Norberg, 2005). However, these are neither reliable nor sensitive for a conclusive diagnosis (Viguier et al., 2009). The ability to readily monitor more reliable mastitis markers on-line with a biosensor during milking would therefore represent a powerful opportunity for the earlier and timely detection of mastitis.

Different sensors for enzymes and proteins, such as NAGase (Pemberton et al., 2001) and Hp (Åkerstedt et al., 2008), have already been developed, and others are on the way. For instance, Mottram and coworkers (Mottram et al., 2007) developed a chemical-array-based sensor that can detect chloride, potassium and sodium ions, together with other inorganic and organic cations and anions. Eriksson and coworkers (Eriksson et al., 2005) developed a gas-sensor-array system, formed by several sensors that detect sulphides, ketones, amines and acids. Again, an increased concentration of lactate does also indicate mastitis already in its early stages (Davis et al., 2004), and a screen-printed sensor based on lactate oxidase has been developed.

Another useful implementation of protein marker measurement consists in the development of rapid, portable 'cow-side' mastitis tests that can enable a more reliable and a less subjective interpretation of results when compared to the CMT or to measuring the EC of milk with hand-held meters, that does not seem to represent a reliable alternative (Pyorala, 2003). Also defined as pen-side, point-of-care (POC) or Rapid Diagnostic Tests (RDTs), these are mostly based on antibody-based techniques, including agglutination, enzyme immunoassays and lateral-flow immunochromatography, and take the form of dipsticks or lateral-flow devices (with the appearance of different symbols or lines depending on the result), latex agglutination systems (coagulation if positive), and in-solution systems (change of solution colour). Usually, cow-side tests do not require dedicated instrumentation for carrying out, reading and interpreting test results, and the reaction occurs in a short time. The main advantage of cow-side tests is that the diagnostic information is readily available where it is needed.

4 Contagious pathogens causing mastitis

4.1 *Streptococcus agalactiae*

Streptococcus agalactiae is a common mastitis agent whose eradication from individual herds is practical and cost-effective. A substantial reduction in milk production and milk quality as measured by SCC is generally associated with infection. *S. agalactiae* infects the cisterns and the ductal system of the mammary gland, causing inflammation of the gland which is mostly subclinical with occasional clinical symptoms.

The majority of infected cows show few clinical signs of mastitis, such as abnormal milk, but generally have high SCC. Mastitis caused by *S. agalactiae* should be suspected in a herd if cow or bulk tank SCCs begin to rise and remain high, especially when bulk milk SCC is 1 000 000 cells/ml or higher. Occasionally high bacteria counts in bulk tank milk will occur when infected udders shed high numbers of *S. agalactiae* in the milk.

Once *S. agalactiae* has been eliminated from a herd, biosecurity measures should be increased and maintained to prevent reinfection. Routine monitoring of bulk tank milk by monthly cultures is a very effective tool for identifying the presence of contagious pathogens in a herd. New infections or outbreaks frequently happen due to the purchase of infected animals. New arrivals should be segregated from the general herd population and milk sampled for culture before joining the milking herd. Dry cows and heifers also need to be included in *S. agalactiae* eradication programmes, since they can represent a source of re-introduction of the organism to the milking herd. Calves fed with milk coming from mothers positive to S. *agalactiae* can spread the infection by suckling themselves or other animals. Once *S. agalactiae* is established within the immature gland, it can persist until first parturition many months later.

4.2 *Staphylococcus aureus*

Staphylococcus aureus is controllable but is more difficult to eradicate than *S. agalactiae.* Infected quarters are the most important source of infection but the organisms can colonize teat skin lesions, the teat canal, and other sites of the cow and eventually pass into the mammary gland. *S. aureus* commonly produces long-lasting infections that can persist through the lactation and into subsequent lactations.

Mastitis caused by *S. aureus* produces more damage to milk-producing tissues than *S. agalactiae*, and decreases milk production with reported losses of 45% per quarter and 15% per infected cow. Farms with bulk tank milk SCC greater than 300 000 to 500 000 cells/mL often have a high prevalence of *S. aureus* infected quarters and a correlated decrease in milk production.

The bacteria damage the duct system and establish deep-seated pockets of infection in the milk secreting tissues, followed by abscess formation and walling-off of bacteria by scar tissue. This walling-off phenomenon is partially responsible for poor cure rates of *S. aureus* infections by antibiotic therapy. During the early stages of infection, damage is minimal and reversible. However, abscesses may release staphylococci to start the infection process in other areas of the gland with further abscess formation and irreversible tissue damage. Occasionally, infection by *S. aureus* may result in peracute mastitis with gangrene. This gangrenous mastitis is characterized by a patchy blue discolouration and coldness of the affected tissue.

S. aureus has also been implicated in intramammary infections in calves, breeding age heifers, and heifers at calving. The source of *S. aureus* to infect these young animals is not known but may be contaminated bedding, feeding milk from *S. aureus* infected cows, cross suckling, or exposure to high fly populations. Pregnant heifers should not be housed together with dry cows, when a significant number of cows in the herd are known to be infected with *S. aureus*.

To prevent *S. aureus* intramammary infections, it is necessary to limit the spread of this organism from cow to cow and to reduce to a minimum the number of infected cows in a herd. To attain these objectives, milk from infected cows should never come in contact with uninfected cows. Since this level of contact is most likely to occur during milking, *S. aureus* infected cows should be identified and milked last, or milked with a separate unit from those used on uninfected cows.

Antibiotic therapy during lactation may improve the clinical condition but usually does not eliminate infection. Infected quarters which do not respond to a single regimen of therapy are generally unresponsive to additional lactation treatment, regardless of culture and antimicrobial sensitivity tests. This is particularly true for mature cows, animals in 3rd lactation or greater. Therapy of 1st lactation animals is likely to provide much high cure rates if accomplished within a month or two of infection. Dry cow therapy may give better results than treatment during lactation, but even then chronic infections can persist into subsequent lactations. *S. aureus* infection status of cows should be one of the considerations when culling decisions are made. Maintaining a *S. aureus*-free herd is possible but more difficult than maintaining a *S. agalactiae*-free herd, and *S. aureus* may reappear even in a closed herd.

4.3 *Mycoplasma* species

Mycoplasma spp. are highly contagious organisms, are less common than *S. agalactiae* and *S. aureus*, and are generally diagnosed in herds experiencing outbreaks of CM that resist therapy. *Mycoplasma* spp. may damage the secretory tissue and produce fibrosis in the udder as well as abscesses with thick fibrous walls, and great enlargement of the supramammary lymph nodes. Frequently, the history of affected herds includes the recent introduction of new animals, a previous outbreak of respiratory disease, and/or cattle with swollen joints. Cattle of all ages and at any stage of lactation are susceptible, but animals in early lactation seem to suffer more severely because of the occurrence of increased mammary gland oedema. In many outbreaks of mycoplasma mastitis it is not uncommon to observe a large portion of infections associated with first-lactation cows.

Mycoplasma spp. should be suspected in herds when multiple cows are unresponsive to treatment, and generally affected cows show a marked drop in the milk production or cease lactating. However, *Mycoplasma* spp. may be isolated from high-producing cows in herds that do not experience the classic signs. Subclinical cases with intermittent signs of CM are not uncommon. Infected cows may have a high SCC and shed organisms for variable periods. A herd suspected of having mycoplasma mastitis, based on history and clinical signs, should be sampled for culture in order to establish the nature of the infection. A significant portion of the herd may be infected but not showing clinical signs of infection. Subclinically infected animals represent a significant risk for ongoing infection in the herd. Mycoplasma mastitis may be complicated by common bacterial infections which occur concurrently.

There is no effective treatment for mycoplasma mastitis, but the disease can be controlled by identifying infected animals by sampling and culturing milk samples from all cows in the herd, followed by segregation and/or culling the infected animals. If *Mycoplasma* spp. infected cows remain in the herd, they should be milked last or with a separate unit from those used on uninfected cows. Great care should be used when purchasing replacements. Many herds become newly infected by adding cows with *Mycoplasma* spp. infected udders. Before commingling with the herd, milk should be cultured from all replacement cows and heifers at calving for *Mycoplasma* spp. as well as for *S. agalactiae* and *S. aureus*. When herds are purchased, it is a good policy to culture all suspected mastitis cows as well as bulk tank milk. When screening bulk tank milk from suspect herds it is recommended that several samplings be evaluated over time.

Sometimes, the disease may suddenly appear in previously uninfected herds without the introduction of replacements. Mycoplasma is widely found as a resident of the bovine upper respiratory tract of apparently normal cows, and transfer of the microorganisms from the respiratory system to the mammary gland can occur. Mycoplasma mastitis outbreaks have been associated with respiratory problems in calves, heifers and cows. Young calves fed milk from cows with *Mycoplasma* spp. infected mammary glands are prone to have respiratory infections and otitis, which may persist for several months.

5 Environmental pathogens: *Escherichia coli*, *Klebsiella* and environmental streptococci

5.1 *Escherichia coli* and *Klebsiella*

In the dairy industry, *E. coli* and *Klebsiella* can be widespread mastitis pathogens in the environment. Acute coliform mastitis is a common and often fatal disease in lactating dairy cows. Endotoxaemia in cows with acute *E. coli* or *Klebsiella* mastitis are generally recognized as the cause of death or culling. Bacteraemia (Wenz et al., 2001) has been reported to occur in 32% (Cebra et al., 1996) to 75% (Katholm and Andersen, 1992) of cows with naturally occurring coliform mastitis. In the United States use of the J-5 core antigen vaccine has greatly reduced morbidity and mortality due to *E. coli* mastitis.

Due to its environmental origin *Klebsiella* mastitis is more difficult to eradicate and treat effectively, causing greater economic losses than other coliform mastitis pathogens such as *E. coli* (Gröhn et al., 2004). *Klebsiella* mastitis can be fatal, and cows that survive infection are often chronically infected with *Klebsiella* mastitis. Chronic cases are generally culled due to persistently high cell counts, recurrent CM and/or reduced milk production.

In a 2011 Cornell University study of 7 NY Holstein herds, researchers looked at the effect of recurrent episodes of CM during different lactations on mortality and culling (Hertl et al., 2011). Infections due to Gram-negative pathogens such as *E. coli* and *Klebsiella* had the greatest impact on milk yield and mortality losses when occurring as a second mastitis case for first calf heifers and as a third case for older cows. Gram-negative mastitis cows were more likely to die than cows with Gram-positive (GP) bacterial mastitis infections after the first two incidences of mastitis. A second study in 2013 found that among first-lactation cows, the presence of a first CM case generally exposed cows to a greater risk of mortality in the current month. The effect of *Klebsiella* infection on individual cows was greater than that of other mastitis pathogens tested, including *E. coli*. The 2nd or 3rd occurrence of clinical *Klebsiella* mastitis resulted in cows in parity ≥2 with an even greater risk of mortality (Cha et al., 2013).

These data suggest that cows infected with *Klebsiella* mastitis may be good candidates for aggressive management including treatment and a more stringent follow-up protocol. Delayed treatment may lead to recurring episodes or chronic infections that are resistant to treatment and/or where increased severity may lead to death. *Klebsiella* and *E. coli* are known to be resistant to the broad-spectrum antibiotic amoxicillin (Roberson et al., 2004), but some success has been achieved with ceftiofur, a broad-spectrum third-generation cephalosporin antibiotic that has been widely used for treatment of mastitis (Oliver and Murinda, 2012). Recent studies have shown that 5-day intramammary use or parenteral administration of ceftiofur was successful in treating non-severe and severe, respectively, CM cases caused by *E. coli* and *Klebsiella* (Schukken et al., 2011; Wenz et al., 2005).

The primary method for control of *E. coli* and *Klebsiella* mastitis continues to be reduction of the pathogen in the environment and reduction of exposure to the animal. Historically, recommendations to lower the incidence and exposure of animals to *E. coli* and *Klebsiella* in the dairy environment have included use of sand bedding, an inorganic material, over other organic material bedding types such as wood shavings and sawdust. A number of studies over the last 10 years have examined the impact of these and other different bedding materials on *E. coli* and *Klebsiella* and other mastitis pathogen counts. With high costs of bedding materials, farmers are looking to different alternatives such as sand, recycled sand or recycled manure solids (RMS). In support of earlier data, more recent research (Harrison et al., 2008; Husfeldt et al., 2012) has found that inorganic bedding materials such as sand had lower initial bacterial counts compared to organic materials such as RMS, recycled paper bedding or wood shavings, particularly when wet. However, when these studies followed the status of bedding material in use in stall areas, it was found that there was little difference between the different bedding types in their ability to harbour high levels of *E. coli* and *Klebsiella* and other pathogens. Faecal shedding of *K. pneumoniae* by a large proportion of dairy cows may explain why *Klebsiella* mastitis occurs in herds that use inorganic bedding material or other bedding material that is free from *Klebsiella* upon introduction into the barn (Munoz et al., 2006). Additionally, recycled sand can serve as a source of *Klebsiella* as organic matter accumulates during the recycled life of this material. Any bedding material can quickly become contaminated in the stall area with mastitis pathogens shed from fresh manure.

5.2 Environmental streptococci

Environmental streptococci and streptococci-like bacteria are significant contributors to the incidence of mastitis (Gröhn et al., 2004) and in the United States 30% or more cows are diagnosed annually with CM caused by these GP, esculin-positive, catalase-negative streptococci. *Streptococcus uberis* as well as a number of Gram-positive cocci have also been associated with bovine mastitis (Jayarao et al., 1991).

S. uberis is a common cause of mastitis in dairy cattle in many countries around the world. *S. uberis* is shed with faeces of cattle and can survive for up to 2 weeks in fresh manure or faeces-contaminated mud or straw (Lopez-Benavides et al., 2007). Generally cows develop intramammary infections if their udders are exposed to contaminated material and especially if they have damaged teat skin or open teat ends. The emergence of *S. uberis* is a problem not only for free stall but also for pasture-based herds with higher stocking rates increasing cow exposure to environmental bacteria. Many cows in the herd can become infected if exposed to environmental bacteria at a vulnerable time: especially in the fortnight after drying-off and the weeks either side of calving or in the hour immediately after milking. Importantly *S. uberis* can also spread from cow-to-cow at milking (Zadoks and Fitzpatrick, 2009).

Research from New Zealand has demonstrated the existence of dominant strains of *S. uberis* which imply that a selection process can occur within the mammary gland, leading to a single strain that is detected upon diagnosis of mastitis within a herd (Pryor et al., 2009). Most quarters that become infected with *S. uberis* have high cell counts (often above 500 000 cells/mL) that return to a normal cell count within 2–3 weeks but a small percentage of cows remain chronically infected and shed bacteria in their milk (Hogan and Smith, 1997). This allows the bacteria to spread from cow-to-cow via the mechanisms associated with contagious mastitis bacteria. So, despite its reputation as an environmental pathogen, control of *S. uberis* also requires attention to management practices that minimize milk droplet 'impacts' associated with vacuum fluctuations within the mechanical milking system.

However, this large and diverse group of Gram-positive cocci also includes other species within *Streptococcus, Enterococcus, Aerococcus* and *Lactococcus* genera. From a clinical perspective it is important to understand the epidemiology of these organisms and the role played to define their behaviour. *Lactococcus lactis* subsp. *lactis* has been recently isolated from bovine mammary glands (Plumed-Ferrer et al., 2015; Plumed-Ferrer et al., 2013; Werner et al., 2014) and bulk tank milk samples (Devriese et al., 1999; Guélat-Brechbuehl et al., 2010) and found to be present in natural soils and plants (Chen et al., 2012; Nomura et al., 2006). The role of *L. lactis* subsp. *lactis* in bovine IMI is still unclear with a few reports of *Lactococcus* isolated in association with bovine and buffalo IMI (Malinowski et al., 2003; Plumed-Ferrer et al., 2013; Pot et al., 1996; Teixeira et al., 1996; Todhunter et al., 1995; Werner et al., 2014).

Mastitis testing laboratories typically do not identify *Streptococcus*-like bacteria to genus or species level beyond identification of *S. agalactiae, S. dygalactiae* and *S. uberis*. Therefore limited data on the isolation of *Lactococcus* species and its impact on udder health exist. Even in laboratories where further identification has been performed, the ability to accurately acquire and report species data on Gram-positive cocci has been confounded by the fact that many routine phenotypic diagnostic tests for streptococci-like bacteria can be inaccurate and unreliable (Odierno et al., 2006). A recent study (Werner et al., 2014) reported that 42 isolates in New York State were phenotypically identified as *S. uberis* and *Streptococcus* spp. but with PCR methods identified 42 isolates (70%) as *L. lactis* subsp. *lactis* .

A number of reports exist documenting discrepancies between the results of different commercially available biochemical identification kits for the identification of streptococci and streptococci-like bacteria from sources including milk (Fortin et al., 2003; Gordoncillo et al., 2010; Svec and Sedlácek, 2008). These authors found a high prevalence of misidentifications when both conventional biochemical and commercial kit testing were performed to identify milk sample streptococci and streptococci-like isolates. Thus, it is possible that the incidence of *L. lactis* subsp. *lactis* in association with bovine IMI has been severely underreported.

In recent years, methods such as MALDI TOF, PCR and sequencing-based methods have proved a more reliable means of accurately differentiating streptococci and in identification of *Lactococcus* spp. even to the subspecies level (Facklam and Elliott, 1995). Using molecular methods, researchers have isolated *L. lactis* subsp. *lactis* and *L. garvieae* from IMI cases, in pure culture (Devriese et al., 1999; Kuang et al., 2009). In some cases molecular methods have resulted in detection of *Lactococcus* spp. in atypical environments. Not normally considered contaminants of water or as a result of inadequate farm hygiene, *L. lactis* subsp. *lactis* was identified by repPCR and ribotyping in the routine analysis of surface waters (Svec and Sedlácek, 2008). These isolates were initially presumptively phenotypically identified as *Enterococcus* spp., bacteria which have often been found in association with surface water. In an investigation of unacceptably high bacterial count bulk milk, Holm et al. (Holm et al., 2004) reported that 32% of samples tested showed growth of *Lactococcus* spp. and 7% of samples were dominated by this type of bacteria. These researchers hypothesized that inadequate milking equipment hygiene and standing water was responsible for the overgrowth of these bacteria. The role of *Lactococci* and other environmental streptococci have not been previously identified, and their prevalence in the microbial environment of dairy farms and as a causal agent of bovine mammary gland infection is still not clear.

6 Other pathogens: *Prototheca*, coagulase-negative staphylococci and other microorganisms

6.1 *Prototheca*

Prototheca spp. is a genus of algae belonging to the family Chlorellaceae. They are ubiquitous in nature, living predominantly in aqueous environments containing decomposing plant material (Anderson and Walker, 1988; Huerre et al., 1993). Within the known *Prototheca* spp., only *P. zopfii*, *P. wickerhamii* and *P. blaschkeae* have been associated with disease in humans and animals (Huerre et al., 1993; Thompson et al., 2009). In humans, protothecosis is mainly caused by *P. wickerhamii* (Lass-Flörl and Mayr, 2007); in veterinary medicine *P. zopfii* is reported as most common causative agent of protothecosis in dogs and cows (Corbellini et al., 2001).

In the past, the genus *Prototheca* was considered a rare pathogen in dairy cattle and associated with infection in the presence of predisposing factors, such as poor environmental conditions and insufficient milking hygiene (Jánosi et al., 2001); however, cases of clinical and chronic mastitis are increasingly recognized and are becoming endemic worldwide (Roesler and Hensel, 2003). Bovine IMIs are most frequently caused by *P. zopfii* infection, whereas *P. wickerhamii* infection is rarely seen. Though there were fewer investigations of buffalo herds with IMI, *Prototheca* spp. infection was more likely related to both *P. zopfii* and *P. wickerhamii*. Almost all *Prototheca* isolates from bovine milk in Italy (Bozzo et al., 2014; Ricchi et al., 2010), Germany (Möller et al., 2007), Portugal (Marques et al., 2008), Poland (Jagielski et al., 2011), Japan (Kishimoto et al., 2010), and China (Gao et al., 2012) were *P. zopfii* genotype 2, suggesting that it is the principal causative agent. However, others have reported the involvement of *P. blaschkeae* in bovine mastitis (Jagielski et al., 2011; Marques et al., 2008; Ricchi et al., 2013).

Diagnosis of *Prototheca* spp. mastitis is typically based on morphological characteristics on culture media. Specialized *Prototheca* Isolation Media (PIM) has been shown to improve diagnosis of *Prototheca* identification. Wet mounts and smears stained with Gram or methylene blue will quickly confirm the diagnosis. Molecular methods (PCR) are available to confirm the species and genotype if necessary. Data recently analysed from an endemically infected herd showed that 24% of *Prototheca*-infected cows (culture positive) had linear scores <4.0 while the remaining 76% of infected cows had a mean linear scores of 5.3 (range 4.0 to 9.6) at the time of diagnosis. Clinical signs of *Prototheca* spp. infection range from watery appearance of milk to palpable swelling, oedema and firmness of the affected quarters. Once the organisms have gained access to the mammary gland *Prototheca* spp. invade macrophages and udder tissue creating a chronic granulomatous lesion (Bozzo et al., 2014; Roesler and Hensel, 2003).

Prototheca spp. are ubiquitous in nature, often found in many locations in the dairy farm environment including water, manure, bedding, forages and other locations associated with high moisture levels and decaying organic matter. *Prototheca* spp. organisms have been found in the faeces of many species of animals including dairy cattle, cats, rats and swine (Anderson and Walker, 1988).

There are no known effective or approved therapies for prototheca mastitis since most infections become chronic with periodic shedding of infective organisms. Recommended management of infected cows includes segregation and culling of culture positive animals. *Prototheca* spp. was originally classified as an environmental, opportunistic mastitis pathogen. However once a critical number of infections is established in the herd, cow-to-cow transmission during milking becomes the dominant cause of new infections. The presence of prototheca organisms in the milking claw and liners after milking an infected cow has been demonstrated, increasing the risk for infection in the next cow to be milked with that unit. In one outbreak where infected cows were housed in the hospital pen it was shown that new prototheca infection risk increased within that pen.

The existence of a large portion of unidentified healthy, subclinically infected shedders in a herd complicates any herd plan to control and eliminate prototheca mastitis from a herd. Wet areas of the cows' environment are often cited as a risk factor for infection as are poor intramammary treatment techniques. Prototheca mastitis can be endemic on farms located in tropical areas. Increased prevalence of infection is seen on farms during periods of warm weather and heavy rainfall (Costa et al., 1997). A recent study in Ontario, Canada (Pieper et al., 2012) revealed that the final logistic regression model for herd-level risk factors included use of intramammary injections of a non-intramammary drug, the number of different injectable antibiotic products being used, the use of any dry cow teat sealant (external or internal), and having treated 3 or more displaced abomasums in the last 12 months. The final logistic regression model for cow-level risk factors included second or greater lactation and the logarithm of the lactation-average SCC. Unsanitary or repeated intramammary infusions, antibiotic treatment, and off-label use of injectable drugs in the udder might promote *Prototheca* udder infection. Risk factors identified by other researchers include increasing parity, antibiotic pretreatment, increased SCC prior to diagnosis of infection and a history of CM (Tenhagen et al., 2001).

6.2 Coagulase-negative staphylococci

Coagulase-negative staphylococci (CNS) are the most prevalent cause of bovine IMI (De Vliegher et al., 2012). Researchers do not define CNS as contagious or environmental pathogens but as 'skin flora opportunists'. Research suggests that CNS are not a homogeneous group (Vanderhaeghen et al., 2015, 2014) and infection rate is higher in cows that have recently given birth. Cows and heifers can have higher prevalence of CNS immediately after calving. *Staphylococcus chromogenes* is the most prevalent species found in IMI while *Staphylococcus equorum*, *Staphylococcus saprophyticus*, *Staphylococcus cohnii* and *Staphylococcus sciuri* are more present in environmental habitats (Piessens et al., 2011) than in milk (Fry et al., 2014), indicating an environmental source. Furthermore, *S. chromogenes*, *S. simulans* and *S. xylosus* have a more substantial effect on udder health than other species. They can cause a considerable increase in quarter SCC (Fry et al., 2014; Supré et al., 2011).

6.3 Other pathogens

Pseudomonas aeruginosa is a bacterium able to cause outbreaks of CM. Generally, a persistent infection occurs, which may be characterized by intermittent acute or subacute exacerbations. The organism is widespread in soil-water environments common to dairy farms. Herd infections have been reported after extensive exposure to contaminated wash water, teat cup liners or intramammary treatments administered by milkers. Failure to use aseptic techniques for udder therapy or use of contaminated milking equipment may lead to establishment of *P. aeruginosa* infections within the mammary glands. Severe peracute mastitis with toxaemia and high mortality may follow immediately in some cows, whereas subclinical infections may occur in others. The organism has persisted in a gland for as long as five lactations, but spontaneous recovery may occur. Other than supportive care for severe episodes, therapy is of little value. Culling is recommended for cows.

Serratia mastitis may arise from contamination of milk hoses, teat dips, water supply or other equipment used in the milking process. The organism is resistant to many disinfectants and antimicrobials. Cows with this form of mastitis that continue to display clinical signs should be culled.

Trueperella pyogenes reclassified from genus *Arcanobacterium* to genus *Trueperella*, is a worldwide known pathogen of domestic ruminants causing mastitis and a variety of pyogenic infections in heifers and dry cows. It is occasionally seen in mastitis of lactating udders after teat injury, and it may be a secondary invader. The inflammation is typified by the formation of profuse, foul-smelling, purulent exudate. Mastitis due to *T. pyogenes* is common among dry cows and heifers that are pastured during the summer months on fields and that have access to ponds or wet areas. The vector for animal-to-animal spread is the fly *Hydrotaea irritans*. Control of infections is by limiting the ability to stand udder-deep in water and by controlling flies. Preventive treatment of heifers and dry cows in susceptible areas with long-acting penicillin preparations has effectively reduced infections. Therapy is rarely successful, and the infected quarter is usually lost to production. Infected cows may be systemically ill, and cows with abscesses usually should be slaughtered.

Nocardia asteroides and other *Nocardia* species are ubiquitous environmental saprophytes, and generally are not a part of the normal flora of mammals, although they may be carried mechanically on the skin. Causes a destructive mastitis characterized by acute onset, high temperature, anorexia and marked swelling of the infected quarter and response in the udder is typical of a granulomatous inflammation. Infection usually arises from inoculation of soft tissue after penetrating injuries, by inhalation of aerosols containing organisms. Slaughter is recommended for infected cows since this organism has zoonotic potential.

Mastitis due to various mycotic organisms (yeasts) has appeared in dairy herds, especially after the use of penicillin in association with prolonged repetitive use of antibiotic infusions in individual cows. Although yeasts are considered opportunistic pathogens, they have been associated with IMI in dairy cattle, commonly related to treatment directed against other pathogens using contaminated syringes, cannulas or contaminated antibiotic preparations (Scaccabarozzi et al., 2011; Spanamberg et al., 2008). The most frequently isolated organisms among mastitis-causing yeasts are *Candida* species, but other yeast genera such as *Trichosporon*, *Cryptococcus* and *Geotrichum* have also been isolated from clinical cases with low frequencies (Chahota et al., 2001; Spanamberg et al., 2008). The susceptibilities of different species to antifungal agents may be different; therefore, no clear evidence exists of the effectiveness of this therapy (Zaragoza et al., 2011). Yeasts grow well in the presence of penicillin and some other antibiotics; they may be introduced during udder infusions of antibiotics, and multiply and cause mastitis.

7 Management and control of mastitis

Routine procedures for the prevention and control of contagious and environmental mastitis have been described in detail by Blowey (2010) and Anonymous (2016a) and will not be reviewed here in any detail. Routine control measures for contagious and environmental mastitis are summarized very briefly below. Routine control measures for mastitis include:

1 Milking routine and hygiene.
2 Maintenance and correct use of milking equipment.
3 Post-milking dipping of teats with disinfectant.
4 Treatment decisions on clinical cases based on severity score and pathogen identification.
5 Dry cow therapy.
6 Culling policy for chronically infected cows.
7 Maintaining a closed herd.

In the case of environmental mastitis, the primary aim is to reduce teat-end exposure to environmental pathogens. This includes management of bedding and walking areas to minimize contamination both in the general housing area and especially in the accommodation around calving.

Environmental conditions that can increase exposure include: overcrowding of pens; poor ventilation; inadequate manure removal from the back of stalls, alleyways, feeding areas and exercise lots; poorly maintained (hollowed out) free stalls; access to farm ponds or muddy exercise lots. Pathogen survival has been found to be longer in dirty, wet conditions (EFSA, 2009; Small et al., 2003).

Bedding materials are a significant source of teat-end exposure to environmental pathogens. The bacteria count in bedding can change depending on contamination with manure, available moisture and temperature. Low-moisture inorganic materials, such as sand or crushed limestone, are preferable to finely chopped organic materials. In general, drier bedding materials are associated with lower numbers of pathogens. Warmer environmental temperatures favour growth of pathogens; lower temperatures tend to reduce growth. Finely chopped organic bedding materials, such as sawdust, shavings, recycled manure, pelleted corncobs, peanut hulls and chopped straw, frequently contain very high coliform and streptococcal numbers. With clean, long straw, coliform numbers are generally low; but the environmental streptococcal numbers may be high. Attempts to maintain low coliform numbers by applying chemical disinfectants or lime are generally impractical because frequent, if not daily, application is required to achieve measurable results. Total daily replacement of organic bedding in the back third of stalls has been shown to reduce exposure of teat ends to coliform bacteria (Buncic, 2006; Vosough Ahmadi et al., 2007).

Post-milking dipping of teats with germicidal dips and ensuring that cows remain standing for ca 30 min post-milking reduces risk by effecting teat closure before exposure of the teats to environmental pathogens. There is evidence that immunizing cows during the dry period with an *E. coli* J-5 bacterin will reduce the severity of clinical coliform cases during early lactation and reduce the recovery time of infected animals (Wilson et al., 2008).

8 Dry cow therapy

Mastitis treatment programmes also need to take into account the different stages in the animal life cycle. As an example, udders can be highly susceptible to infection during the early dry period. However, the dry period is also an important opportunity to rid the udder of potential pathogens that cause mastitis. The majority of new intramammary infections are most likely to be subclinical during the dry period and these infections can then become clinical in early lactation (Green et al., 2002). It has been estimated that 55% of environmental infections established early in the dry period, including Gram-negative intramammary infections, can persist into the next lactation and can result in CM cases (Todhunter et al., 1995). In fact, 52% of all clinical coliform mastitis cases occurring in the first 100 DIM of lactation may originate during the previous dry period (Bradley and Green, 2000). Smith et al. (1985) also reported that the risk for new intramammary infections from environmental pathogens can be 10 times higher during the dry period than during lactation.

There are two periods of elevated risk for new infections during gestation as demonstrated by Fig. 1. The first is immediately after dry off, prior to involution of the udder. The second period of increased risk extends from approximately three weeks prior to calving (colostrogenesis) until approximately three weeks after calving. The risk for new infection between these two periods, once involution of the udder is complete, is minimal.

The primary objective of dry cow therapy is to cure susceptible existing subclinical infections present at dry off, such as *S. aureus* and streptococci, and to prevent new intramammary infections which could be acquired during the early dry period. Dry cow therapy has helped to reduce the new intramammary infections risk from 30% to 60% in untreated cows down to 0 to 15% in treated cows (Halasa et al., 2009) and it has also been associated with a decrease in CM cases during lactation (Whist et al., 2006). Antimicrobial dry cow therapy has been found to be effective for approximately two to five weeks after

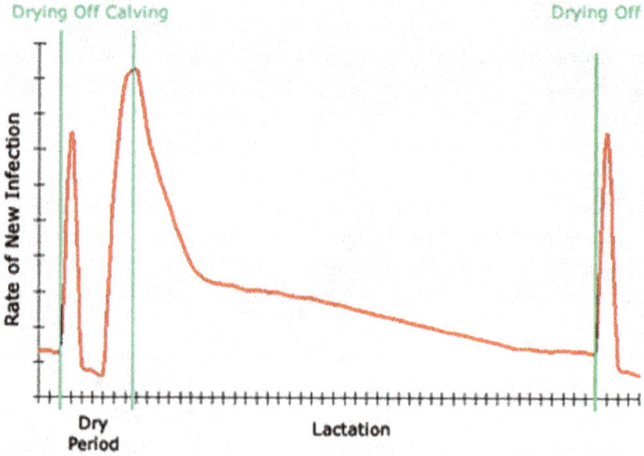

Figure 1 Relative rate of new intramammary infection throughout gestation in the cow. Source: Bradley et al. (2004).

administration. It has been found that dry cow therapy did not prevent new IMI during the latter part of the dry period especially in animals with extended dry periods (Robert et al., 2005; Smith et al., 1967; Oliver et al., 1992).

Good therapeutic practice includes treating all quarters of each cow with antibiotic products specifically designed for dry cow therapy. Most dry treatments are designed to eliminate or at least reduce existing infections by Gram-positive bacteria, particularly *S. aureus* and streptococcal infections at drying-off (Timms, 2000). On many farms, especially those where dairy cattle confinement has become more intense, a higher percentage of new infections during the dry period are caused by environmental bacteria. Most products are reasonably effective against environmental streptococci, especially *S. uberis*, but lack activity against Gram-negative environmental bacteria, especially the coliforms.

It has been shown that these treatments are best used alongside other techniques such as teat sealants (Berry and Hillerton, 2002; Dingwell et al., 2003; Green et al., 2007; Bradley et al., 2011). Use of external and internal teat sealants has been suggested to prevent infection (Godden et al., 2003). Internal sealants in particular have been shown to be effective in reducing the incidence of new infections during the dry period (Cook et al., 2004; Crispie et al., 2004). The keratin plug has been seen as an important method of protection against IMI and it has been shown that the risk of IMI increases in cows with impaired plug integrity (Bramley and Dodd, 1984; Capuco et al., 1992). Quarters that formed a keratin plug early in the dry period had a lower risk of infection than those that did not close (10% and 14% respectively) (Dingwell et al., 2004). The use of an internal teat sealant also provides added protection from new infections during the last few weeks of the dry period.

Blanket dry cow therapy (BDCT), treatment of all cows immediately after their last milking, has been recommended since the 1960s. Treatment of all cows at dry off reaches all cows and is more effective in preventing new infections during the dry period than selective dry cow therapy (SDCT). BDCT does not require the additional expense of diagnostic tests or historical cow data at dry off to determine which cows may benefit from selective therapy.

SDCT has received increasing attention in recent years owing to global concerns over agricultural use of antimicrobial drugs and development of antimicrobial resistance to medically important antimicrobials. When subclinical mastitis in a herd has been reduced to a very low level (e.g. every cow in the herd less than 100 000 cells/ml), using dry cow treatment only on selected higher risk cows has been considered appropriate by some dairy producers and veterinarians. Most studies indicate that if the decision is based on economics (i.e. the cost of dry cow therapy compared to the return to the producer), treating every quarter of every cow at drying-off remains more attractive.

Strategies for instituting a successful SDCT programme have yet to be developed to provide reasonable control of new intramammary infections and a significant reduction in dry cow therapy antimicrobial use. Selection of animals for therapy with dry cow antibiotics based on their SCC at the last milk recording before drying-off gives a substantial reduction in antibiotic use, but risks an increase in subclinical mastitis (Scherpenzeel et al., 2014). A study of 16 commercial herds in Canada was conducted to determine the effect of a Petri film on-farm culture-based selective DCT programme on milk yield and SCC in the following lactation (Cameron et al., 2015). When low-SCC cows were selectively treated with DCT at drying-off based on results obtained using the Petri film on-farm culture system, no effect on milk production for blanket DCT vs. selective DCT or linear score was observed in the subsequent lactation when compared with cow receiving blanket DCT. The results of this study indicate that selective DCT based on results obtained by the

Petri film on-farm culture system enabled a reduction in the use of DCT without negatively affecting milk production and milk quality.

Most studies indicate that when the decision to develop a strategy for management of mastitis infection risk (BDCT vs. SDCT) is based on economics, treatment of all quarters of all cows at dry off is still attractive. A simulation study comparing the dry period intervention studies including BDCT (the default scenario), BDCT combined with teat sealant (TS), SDCT or TS, and SDCT combined with TS showed that BDCT had the lowest combined total annual net cost of IMI; however, the differences among the four scenarios were minor. The simulation results also showed that a considerable number of cows acquire new IMI during the dry period and start the new lactation with IMI. Analysis by Bradley et al. (2015) also suggested that the early and mid-dry period may be more important with respect to the timing of acquisition of new infection than previously thought. They observed substantial variation in the aetiology and prevalence of different pathogens on different farms. This study highlights the importance of assessing and understanding infection dynamics on individual units. The lack of influence of the cow and quarter factors measured in this study suggests that herd and management factors may be more influential.

As has been noted, use of dry cow therapy is one component of an effective mastitis control programme that should also include: proper milking procedures using properly functioning milking equipment, dipping teats immediately after milking with a product known to be safe and effective, good udder hygiene between milkings, keeping accurate records of CM and SCCs on individual cows to assist in making management decisions, treating all clinical cases of mastitis promptly and appropriately, and culling cows with chronic mastitis.

9 The use of antibiotics

The debate about the relative merits of blanket dry cow therapy (BDCT) and selective dry cow therapy (SDCT) highlights the broader questions about the role of antibiotics in animal health. Ensuring animal health includes components such as an effective herd health management programme involving steps such as regularly checking animals for disease, isolating sick animals, providing appropriate treatment and, where necessary, culling chronically infected animals. A dairy herd health plan can also include specific measures, such as the use of antibiotics and vaccines, to protect animals from infection with diseases such as mastitis (Erskine, 2000; Hillerton and Berry, 2005; Suojala et al., 2013). Vaccines are available to prevent several important bacterial mastitis pathogens. The core antigen *E coli* vaccine, for example, has been demonstrated to substantially reduce the severity of coliform mastitis and reduce the need for intramammary treatment (DeGraves et al., 1991).

However, the efficacy of antibiotic therapy varies, often because antibiotics do not penetrate the infected area and have poor intracellular penetration (Pyorala, 2009). Many infections caused by streptococci and staphylococcal species respond well to appropriate treatment with cure rates of 80% or higher. Treatment success of infections caused by other organisms including some coliform organisms (*E. coli* and *Klebsiella* species) may respond to a few select antimicrobials (Schukken et al., 2011). Most other Gram-negative species respond poorly to IMM treatment and are managed by means other than antibiotic therapy. Mastitis caused by other pathogens such as yeast is exacerbated by IMM treatment and some, like prototheca species and mycoplasma, have no known treatment (Lago et al., 2011a; Fuenzalida et al., 2016). There is thus strong evidence that therapy does not always imply biological cure of infected animals or a significant reduction of SCC of the treated quarter. Healing of cows may not imply a financial advantage for farmers. Under these conditions, therapy of mastitis mainly aims at obtaining productive efficiency at a reasonable cost.

Effective use of antibiotics depends on sampling to identify the bacteria and reviewing the cow's medical history as well as treatment outcomes for previous cases on the farm in order to evaluate the chances of therapy being effective, with a focus on using narrow spectrum antibiotics for as short a period as possible (Lago et al., 2011; Ruegg, 2011; Oliveira and Ruegg, 2014). It is important that managers and herd veterinarians look at other individual cow characteristics when making management decisions. These should include parity, days-in-milk, reproductive status (days carried calf), milk production, other health issues, and previous mastitis pathogen identification. In many cases, isolating sick animals may be more effective in disease control. Some cows may be obvious cull candidates; others may be classified as 'do-not-breed' (DNB) for future cull planning, and therefore are not targeted for a milk culture. Animals designated as DNB will remain in the herd until they are unproductive, at which point they are culled.

Other treatment options include the use of bacteriophages, viruses which infect pathogens producing mastitis (Gill et al., 2006). Mineral and vitamin supplementation (e.g. selenium and vitamin E) can also help boost immune function to prevent mastitis (Weiss, 2002). Good nutrition and feeding management of the dry and transition period of gestation/lactation is seen as critically important for maintaining the immune system competence of the cow in early lactation (Lacetera et al., 2005; Sordillo et al., 2009; Trevisi et al., 2011). Today immunomodulators are being introduced therapeutically as means of managing cows' immune response and should reduce the need for antimicrobial therapy as these therapies are developed.

The treatment decision process varies greatly from farm to farm as well as with veterinary practitioners. On many farms, treatment of CM with intramammary (IMM) medication can be substantially reduced without compromising treatment success by means of a pathogen-based treatment protocol. The pathogen-based treatment decision strategy currently used by many veterinarians is to base all IMM treatment judgements on the culture result of milk from the inflamed quarter and severity of clinical signs. Typically the severity of the inflammatory reaction can be characterized in two or three categories:

- Category I cases are characterized by visually abnormal milk.
- Category II cases will include visually abnormal milk as well as a swollen or painful udder. Some veterinarians combine Categories I and II into a single 'mild' class.
- Category III (severe) cases include cows with systemic signs of illness. Animals with severe cases of clinical mastitis are characterized by fever, dehydration, depression and loss of appetite.

The most common bottleneck encountered when establishing a pathogen-based treatment programme is timely receipt of the culture result at the farm. A competent practice-based mastitis diagnostic facility or on-farm lab is the suitable solution in many situations. On-farm diagnostics can provide relatively simple results such as Gram-positive (GP), Gram-negative (GN) or no growth that can direct treatment selection and reduce antibiotic use.

Lago et al. (2011a,b) compared the use of a selective treatment programme based on On-Farm Culture (OFC) results (only GP cases received IMM antibiotics) and a treatment programme in which all cases received IMM antibiotics. No significant differences in short term or long term health and performance outcomes were found. However, drug use was significantly reduced, with only 44% of cows assigned to the OFC system receiving IMM therapy. Another study using Petri films as the diagnostic method conducted in 48 Canadian dairy herds reported a 40% decrease in drug use by using OFC to guide strategic treatment of only GP cases. They concluded that cure rates and long-term health risks were not adversely affected in accurately diagnosed cases (MacDonald et al., 2011).

Vasquez et al. (2016) reported on a treatment trial conducted at a large dairy herd in New York. Using a randomized design, cows with CM scores (CS) of 1 or 2 (signs limited to abnormal milk and/or a swollen or painful udder were assigned to either the blanket therapy (BT) or culture-based therapy (CBT) group. Cows with a CS of 3 were excluded from the trial (systemically ill). Samples were retrieved daily and results were available after 24 hours by direct electronic upload onto farm computers.

Cows in the blanket therapy (BT) group received the current clinical mastitis protocol used by the farm. Cows assigned to the culture group (CBT) received no treatment for the first 24 h. Upon upload of results, the following treatment protocol was automatically assigned by the farm management software for the relevant animal: *Staphylococcus* spp., *Streptococcus* spp. or *Enterococcus* were given an IMM tube of cephapirin sodium once every 12 h for 2 treatments. Cows positive for other organisms or no growth received no treatment.

No statistically significant differences were found between blanket therapy and CBT cows in days to clinical cure. No statistical differences were observed in next test day milk production between groups. Risk of culling before 30 days post enrolment was also statistically the same for both groups, as was risk of culling prior to 60 days. The resulting decrease in antibiotic costs resulted in a significantly increased cash flow.

A similar treatment decision strategy can be applied to the management and treatment of chronic subclinical mastitis to reduce chronic infections in the herd and manage the bulk tank somatic cell count BTSCC. The vast majority of somatic cells found in bulk tank milk (>70%) most often originate from cows with subclinical mastitis. Although many commercially available IMM products are labelled for use in subclinically infected animals, they are under-utilized as a tool to manage chronic subclinical infections. Their use to manage chronic subclinical mastitis and bulk tank milk SCC is seeing a resurgence of interest. This approach does not compromise treatment success, yet can significantly reduce treatment costs, control and reduce the flow of animals through the treatment pens, substantially reduce discarded milk, and reduce drug use and the risk for drug residues.

10 Where to look for further information

- Canadian Bovine Mastitis Research Network (http://www.medvet.umontreal.ca/reseau_mammite/en/index.php)
- CellCheck (http://animalhealthireland.ie/)
- DairyCo Mastitis Control Plan (http://www.mastitiscontrolplan.co.uk/)
- Dutch Udder Health Centre (GD) (http://www.gdanimalhealth.com/)
- M-Team, University of Ghent (http://www.m-team.ugent.be/v2/home/)
- National Mastitis Council (NMC) (https://www.nmconline.org/)

- New York State Cattle Health Assurance Program – Milk Quality Module
- (https://ahdc.vet.cornell.edu/programs/NYSCHAP/modules/mastitis/index.cfm)
- Quality Milk Production Services (QMPS), Cornell University (https://ahdc.vet.cornell.edu/sects/QMPS/)
- Southeast Quality Milk Initiative (http://sequalitymilk.com/)
- University of Minnesota Udder Health Laboratory (https://www.vdl.umn.edu/services-fees/udder-health-mastitis)
- University of Wisconsin Milk Quality website (http://milkquality.wisc.edu/)

Reference works:

- *Rebhun's Diseases of Dairy Cattle*, 2nd Edition is the most recent guide to dairy cattle disease management. This volume contains the most comprehensive coverage of diseases and medical management of udder health issues. The book is organized by body system for quick, convenient reference, and this new edition meets the growing need for management of both diseases of individual cows and problems affecting whole herds. A third volume is due out in 2017, by Thomas J. Divers, DVM, Dipl ACVIM, ACVECC and Simon Peek, BVSc, MRCVS, PhD, Diplomate ACVIM, ISBN: 9781416031376.
- *Veterinary Clinics of North America – Food Animal Practice*, 28(2), July 2012; 19(1), March 2003.
- *Dairy Freestall Housing and Equipment*, MidWest Plan Service, Iowa State University, Ames, Iowa, 8th Edition 2013, by B. Holmes, N. Cook, T. Funk, R. Graves, D. Kammel, D. J. Reinemann and J. Zulovich, Laboratory Handbook on Bovine Mastitis, 2017, 3rd Edition, NMC edition.
- The Laboratory Handbook on Bovine Mastitis includes information, photographs and illustrations, regarding sample collection and handling, diagnostic equipment and materials, diagnostic procedures, molecular diagnostics, mastitis pathogens, somatic cell count, bulk tank culture and on-farm culture. Also, the book contains appendices on general isolation media, mycoplasma medium and testing procedures, other media, testing procedures and stains.

11 References

Abd El-Salam, M. H., 2014. Application of Proteomics to the areas of milk production, processing and quality control – A review. *Int. J. Dairy Technol.* 67, n/a–n/a. doi:10.1111/1471-0307.12116.

Addis, M. F., Pisanu, S., Ghisaura, S., Pagnozzi, D., Marogna, G., Tanca, A., Biosa, G., Cacciotto, C., Alberti, A., Pittau, M., Roggio, T. and Uzzau, S., 2011. Proteomics and pathway analyses of the milk fat globule in sheep naturally infected by Mycoplasma agalactiae provide indications of the in vivo response of the mammary epithelium to bacterial infection. *Infect. Immun.* 79, 3833–45. doi:10.1128/IAI.00040-11.

Addis, M. F., Pisanu, S., Marogna, G., Cubeddu, T., Pagnozzi, D., Cacciotto, C., Campesi, F., Schianchi, G., Rocca, S. and Uzzau, S., 2013. Production and release of antimicrobial and immune defense proteins by mammary epithelial cells following Streptococcus uberis infection of sheep. *Infect. Immun.* 81, 3182–97. doi:10.1128/IAI.00291-13

Addis, M. F., Tedde, V., Dore, S., Pisanu, S., Puggioni, G. M. G., Roggio, A. M., Pagnozzi, D., Lollai, S., Cannas, E. A. and, Uzzau, S., 2016a. Evaluation of milk cathelicidin for detection of dairy sheep mastitis. *J. Dairy Sci.* 99, 6446–56. doi: 10.3168/jds.2015-10293.

Addis, M. F., Tedde, V., Puggioni, G. M. G., Pisanu, S., Casula, A., Locatelli, C., Rota, N., Bronzo, V., Moroni, P. and Uzzau, S., 2016b. Evaluation of milk cathelicidin for detection of bovine mastitis. *J. Dairy Sci.* 99, 8250–8. doi: 10.3168/jds.2016-11407.

Åkerstedt, M., Waller, K. P., Larsen, L. B., Forsbäck, L. and Sternesjö, Å., 2008. Relationship between haptoglobin and serum amyloid A in milk and milk quality. *Int. Dairy J.* 18, 669–74. doi:10.1016/j.idairyj.2008.01.002

Allen, J. C., 1990. Milk synthesis and secretion rates in cows with milk composition changed by oxytocin. *J. Dairy Sci.* 73, 975–84. doi:10.3168/jds.S0022-0302(90)78755-3

Anderson, K. L. and, Walker, R. L., 1988. Sources of Prototheca spp in a dairy herd environment. *J. Am. Vet. Med. Assoc.* 193, 553–6.

Anonymous 2016a. *Current Concepts of Bovine Mastitis 5th Edition*, National Mastitis Council (NMC), New Prague – Minnesota, USA.

Anonymous 2016b. Milk: *Hygiene on the Dairy Farm: A Practical Guide for Milk Producers*, Food Standards Agency, London, UK (https://www.food.gov.uk/business-industry/farmingfood/dairy-guidance/milk-hygiene-guide-for-milk-producers).

Auldist, M. J., Coats, S., Rogers, G. L. and McDowell, G. H., 1995. Changes in the composition of milk from healthy and mastitic dairy cows during the lactation cycle. *Aust. J. Exp. Agric.* 35, 427–36.

Auldist, M. J., Coats, S., Sutherland, B. J., Mayes, J. J., McDowell, G. H. and Rogers, G. L., 1996. Effects of somatic cell count and stage of lactation on raw milk composition and the yield and quality of Cheddar cheese. *J. Dairy Res.* 63, 269–80.

Auldist, M. J. and Hubble, I. B., 1998. Effects of mastitis on raw milk and dairy products. *Aust. J. Dairy Technol.* 53, 28–36.

Babaei, H., Mansouri-Najand, L., Molaei, M. M., Kheradmand, A. and Sharifan, M., 2007. Assessment of lactate dehydrogenase, alkaline phosphatase and aspartate aminotransferase activities in cow's milk as an indicator of subclinical mastitis. *Vet. Res. Commun.* 31, 419–25. doi:10.1007/s11259-007-3539-x

Bar, D., Gröhn, Y. T., Bennett, G., González, R. N., Hertl, J. A., Schulte, H. F., Tauer, L. W., Welcome, F. L. and Schukken, Y. H., 2007. Effect of repeated episodes of generic clinical mastitis on milk yield in dairy cows. *J. Dairy Sci.* 90, 4643–53. doi:10.3168/jds.2007-0145.

Barbano, D. M., Ma, Y. and Santos, M. V, 2006. Influence of raw milk quality on fluid milk shelf life. *J. Dairy Sci.* 89 Suppl 1, E15–9. doi:10.3168/jds.S0022-0302(06)72360-8.

Barkema, H., Green, M., Bradley, A. and Zadoks, R., 2009. The role of contagious disease in udder health, *J. Dairy Sci.* 92 (10): 4717–29.

Barkema, H. W., Schukken, Y. H., Lam, T. J., Galligan, D. T., Beiboer, M. L. and Brand, A., 1997. Estimation of interdependence among quarters of the bovine udder with subclinical mastitis and implications for analysis. *J. Dairy Sci.* 80, 1592–9. doi:10.3168/jds.S0022-0302(97)76089-2.

Berry, E. and Hillerton, J., 2002. The effect of selective dry cow treatment on new intramammary infections, *J. Dairy Sci.* 85 (1): 112–21.

Blowey, R., 2010. *Mastitis Control in Dairy Herds*, CABI Publishers, Wallingford, UK.

Bogin, E. and Ziv, G., 1973. Enzymes and minerals in normal and mastitic milk. *Cornell Vet.* 63, 666–76.

Bozzo, G., Bonerba, E., Di Pinto, A., Bolzoni, G., Ceci, E., Mottola, A., Tantillo, G. and Terio, V., 2014. Occurrence of Prototheca spp. in cow milk samples. *New Microbiol.* 37, 459–64.

Bradley, A. J. and Green, M. J., 2000. A study of the incidence and significance of intramammary enterobacterial infections acquired during the dry period. *J. Dairy Sci.* September 83 (9): 1957–65. PubMed PMID: 11003224.

Bradley, A. J. and Green, M. J., 2004. The importance of the nonlactating period in the epidemiology of intramammary infection and strategies for prevention. *Vet. Clin. North Am. Food Anim. Pract.* 20 (3): 547–68. Review. PubMed PMID: 15471624.

Bradley, A. and Green, M., 2005. Use and interpretation of somatic cell count data in dairy cows. *In Pract.* 27, 310–15. doi:10.1136/inpract.27.6.310.

Bradley, A., Breen, J., Payne, B. and Green, M., 2011. A comparison of broad-spectrum and narrow-spectrum dry cow therapy used alone and in combination with a teat sealant, *J. Dairy Sci.* 94 (2): 692–704.

Bramley, A. J. and Dodd, F. H., 1984. Reviews of the progress of dairy science: mastitis control – progress and prospects. *J. Dairy Res.* August 51 (3): 481–512. Review. PubMed PMID: 6381562.

Brennan, M. and Christley, R., 2012. Biosecurity on cattle farms: a study in North-West England, *PLoS One* 7 (1). doi: 10.1371/journal.pone.0028139.

Bruckmaier, R. M., Ontsouka, C. E. and Blum, J. W., 2004. Fractionized milk composition in dairy cows with subclinical mastitis. *Vet. Med.* – UZPI (Czech Republic) 49, 283–90.

Buncic, S., 2006. *Integrated Food Safety and Veterinary Public Health*, CABI Publishers, Wallingford, UK.

Cameron, M., Keefe, G. P., Roy, J. P., Stryhn, H., Dohoo, I. R. and McKenna, S. L., 2015. Evaluation of selective dry cow treatment following on-farm culture: milk yield and somatic cell count in the subsequent lactation. *J. Dairy Sci.* April 98 (4): 2427–36. doi: 10.3168/jds.2014-8876. PubMed PMID: 25648799.

Capuco, A. V., Bright, S. A., Pankey, J. W., Wood, D. L., Miller, R. H. and Bitman, J., 1992. Increased susceptibility to intramammary infection following removal of teat canal keratin. *J. Dairy Sci.* 75 (8): 2126–30. PubMed PMID: 1383301.

Cebra, C. K., Garry, F. B. and Dinsmore, R. P., 1996. Naturally occurring acute coliform mastitis in holstein cattle. *J. Vet. Intern. Med.* 10, 252–7. doi:10.1111/j.1939-1676.1996.tb02058.x.

Ceciliani, F., Ceron, J. J., Eckersall, P. D. and Sauerwein, H., 2012. Acute phase proteins in ruminants. *J. Proteomics* 75, 4207–31. doi:10.1016/j.jprot.2012.04.004.

Ceciliani, F., Eckersall, D., Burchmore, R. and Lecchi, C., 2014. Proteomics in veterinary medicine: applications and trends in disease pathogenesis and diagnostics. *Vet. Pathol.* 51, 351–62. doi:10.1177/0300985813502819.

Cha, E., Bar, D., Hertl, J. A., Tauer, L. W., Bennett, G., González, R. N.,Schukken, Y. H., Welcome, F. L. and Gröhn, Y. T., 2011. The cost and management of different types of clinical mastitis in dairy cows estimated by dynamic programming. *J. Dairy Sci.* 94: 4476–87. doi: 10.3168/jds.2010-4123.

Cha, E., Hertl, J. A., Schukken, Y. H., Tauer, L. W., Welcome, F. L. and Gröhn, Y. T., 2013. The effect of repeated episodes of bacteria-specific clinical mastitis on mortality and culling in Holstein dairy cows. *J. Dairy Sci.* 96, 4993–5007. doi:10.3168/jds.2012-6232.

Chagunda, M. G., Larsen, T., Bjerring, M. and Ingvartsen, K. L., 2006. L-lactate dehydrogenase and N-acetyl-beta-D-glucosaminidase activities in bovine milk as indicators of non-specific mastitis. *J. Dairy Res.* 73, 431–40. doi:10.1017/S0022029906001956.

Chahota, R., Katoch, R., Mahajan, A. and Verma, S., 2001. Clinical bovine mastitis caused by Geotrichum candidum. *Vet. Ark.* 71, 197–201.

Chen, M.-H., Hung, S.-W., Shyu, C.-L., Lin, C.-C., Liu, P.-C., Chang, C.-H., Shia, W.-Y., Cheng, C.-F., Lin, S.-L., Tu, C.-Y., Lin, Y.-H. and Wang, W.-S., 2012. Lactococcus lactis subsp. lactis infection in Bester sturgeon, a cultured hybrid of Huso huso × Acipenser ruthenus, in Taiwan. *Res. Vet. Sci.* 93, 581–8. doi:10.1016/j.rvsc.2011.10.007.

Cook, N. B., Bennett, T. B., Emery, K. M. and Nordlund, K. V, 2002. Monitoring nonlactating cow intramammary infection dynamics using DHI somatic cell count data. *J. Dairy Sci.* 85, 1119–26. doi:10.3168/jds.S0022-0302(02)74173-8.

Cook, N., Wilkinson, A., Gajewski, K., Weigel, D. and Sharp, P., 2004. The prevention of new intramammary infections during the dry period when using an internal teat sealant in conjunction with a dry cow antibiotic, Proceedings of the National Mastitis Council, Charlotte – North Carolina, USA.

Corbellini, L. G., Driemeier, D., Cruz, C., Dias, M. M. and Ferreiro, L., 2001. Bovine mastitis due to Prototheca zopfii: clinical, epidemiological and pathological aspects in a Brazilian dairy herd. *Trop. Anim. Health Prod.* 33, 463–70.

Costa, E. O., Melville, P. A., Ribeiro, A. R., Watanabe, E. T. and Parolari, M. C., 1997. Epidemiologic study of environmental sources in a Prototheca zopfii outbreak of bovine mastitis. *Mycopathologia* 137, 33–6.

Crispie, F., Flynn, J., Ross, R., Hill, C. and Meaney, W., 2004. Dry cow therapy with a non-antibiotic intramammary teat seal: a review, *Ir. Vet. J.* 57 (7): 412–18.

Davis, S. R., Farr, V. C., Prosser, C. G., Nicholas, G. D., Turner, S.-A., Lee, J. and Hart, A. L., 2004. Milk L-lactate concentration is increased during mastitis. *J. Dairy Res.* 71, 175–81. doi:10.1017/S002202990400007X.

De Vliegher, S., Fox, L. K., Piepers, S., McDougall, S. and Barkema, H. W., 2012. Invited review: Mastitis in dairy heifers: nature of the disease, potential impact, prevention, and control. *J. Dairy Sci.* 95, 1025–40. doi:10.3168/jds.2010-4074.

DeGraves, F. J. and Fetrow, J., 1991. Partial budget analysis of vaccinating dairy cattle against coliform mastitis with an *Escherichia coli* J5 vaccine. *J. Am. Vet. Med. Assoc.* August 15, 199 (4): 451–5. PubMed PMID: 1917656.

Deluyker, H. A., Van Oye, S. N. and Boucher, J. F., 2005. Factors affecting cure and somatic cell count after pirlimycin treatment of subclinical mastitis in lactating cows. *J. Dairy Sci.* 88, 604–14. doi:10.3168/jds.S0022-0302(05)72724-7.

Devriese, L. A., Hommez, J., Laevens, H., Pot, B., Vandamme, P. and Haesebrouck, F., 1999. Identification of aesculin-hydrolyzing streptococci, lactococci, aerococci and enterococci from subclinical intramammary infections in dairy cows. *Vet. Microbiol.* 70, 87–94.

Devriese, L. A., Schleifer, K. H. and Adegoke, G. O., 1985. Identification of coagulase-negative staphylococci from farm animals. *J. Appl. Bacteriol.* 58, 45–55.

Dingwell, R., Kelton, D. and Leslie, K., 2003. Management of the dry cow in control of peripartum disease and mastitis, *Vet. Clin. North Am. Anim. Prac.* 19 (1): 235–65.

Dingwell, R. T., Leslie, K. E., Schukken, Y. H., Sargeant, J. M., Timms, L. L., Duffield, T. F., Keefe, G. P., Kelton, D. F., Lissemore, K. D. and Conklin, J., 2004. Association of cow and quarter-level factors at drying-off with new intramammary infections during the dry period. *Prev. Vet. Med.* April 30, 63(1–2): 75–89. PubMed PMID: 15099718.

Djabri, B., Bareille, N., Beaudeau, F. and Seegers, H., 2002. Quarter milk somatic cell count in infected dairy cows: A meta-analysis. *Vet. Res.* 33, 335–57. doi:10.1051/vetres:2002021.

Dohoo, I. R. and Leslie, K. E., 1991. Evaluation of changes in somatic cell counts as indicators of new intramammary infections. *Prev. Vet. Med.* 10, 225–37. doi:10.1016/0167-5877(91)90006-N.

Dohoo, I. R. and Meek, A. H., 1982. Somatic cell counts in bovine milk. *Can. Vet. J. La Rev. vétérinaire Can.* 23, 119–25.

Eaton, J. W., Brandt, P., Mahoney, J. R. and Lee, J. T., 1982. Haptoglobin: a natural bacteriostat. *Science* 215, 691–3.

Eckersall, P. D., Young, F. J., McComb, C., Hogarth, C. J., Safi, S., Weber, A., McDonald, T., Nolan, A. M. and Fitzpatrick, J. L., 2001. Acute phase proteins in serum and milk from dairy cows with clinical mastitis. *Vet. Rec.* 148, 35–41.

EFSA. 2009. Scientific Opinion of the panel on biological Hazards on a request from the European Commission on Food Safety Aspects of Dairy Cow Housing and Husbandry Systems, *EFSA J.* 1189: 1–27.

Eriksson, Å., Persson Waller, K., Svennersten-Sjaunja, K., Haugen, J.-E., Lundby, F. and Lind, O., 2005. Detection of mastitic milk using a gas-sensor array system (electronic nose). *Int. Dairy J.* 15, 1193–201. doi:10.1016/j.idairyj.2004.12.012.

Erskine, R., 2000. Antimicrobial drug use in bovine mastitis, in Prescott, J., Baggot, J. and Walker, R. (eds), *Antimicrobial Therapy in Veterinary Medicine*, Iowa State University Press, Ames, USA.

Facklam, R. and Elliott, J. A., 1995. Identification, classification, and clinical relevance of catalase-negative, gram-positive cocci, excluding the streptococci and enterococci. *Clin. Microbiol. Rev.* 8, 479–95.

Fernandes, A. M., Oliveira, C. A. F. and Lima, C. G., 2007. Effects of somatic cell counts in milk on physical and chemical characteristics of yoghurt. *Int. Dairy J.* 17, 111–15. doi:10.1016/j.idairyj.2006.02.005.

Forsbäck, L., Lindmark-Månsson, H., Andrén, A. and Svennersten-Sjaunja, K., 2010. Evaluation of quality changes in udder quarter milk from cows with low-to-moderate somatic cell counts. *Animal* 4, 617–26. doi:10.1017/S1751731109991467.

Fortin, M., Messier, S., Paré, J. and Higgins, R., 2003. Identification of catalase-negative, non-beta-hemolytic, gram-positive cocci isolated from milk samples. *J. Clin. Microbiol.* 41, 106–9.

Fry, P. R., Middleton, J. R., Dufour, S., Perry, J., Scholl, D. and Dohoo, I., 2014. Association of coagulase-negative staphylococcal species, mammary quarter milk somatic cell count, and persistence of intramammary infection in dairy cattle. *J. Dairy Sci.* 97, 4876–85. doi:10.3168/jds.2013-7657.

Fuenzalida, et al., 2006. Preliminary results of an ongoing clinical trial evaluating effects of treatment of culture-negative cases of clinical mastitis on SCC and milk production. NMC Annual Meeting Proceedings.

Gao, J., Zhang, H., He, J., He, Y., Li, S., Hou, R., Wu, Q., Gao, Y. and Han, B., 2012. Characterization of Prototheca zopfii associated with outbreak of bovine clinical mastitis in herd of Beijing, China. *Mycopathologia* 173, 275–81. doi:10.1007/s11046-011-9510-y.

Gill, J. et al., 2006. Efficacy and pharmacokinetics of bacteriophage therapy in treatment of subclinical *Staphylococcus aureus* mastitis in lactating dairy cattle, *Antimicrob. Agents Chemother.* 50 (9): 2912–18.

Godden, S., Rapnicki, P., Stewart, S., Fetrow, J. Johnson, A., Bey, R. and Farnsworth, R., 2003. Effectiveness of an internal teat sealant in the prevention of new intra-mammary infections during the dry and early lactation periods in dairy cows when used with a dry cow intra-mammary antibiotic, *J. Dairy Sci.* 86, 3899–11.

Gordoncillo, M. J. N., Bautista, J. A. N., Hikiba, M., Sarmago, I. G. and Haguingan, J. M. B., 2010. Comparison of conventionally identified mastitis bacterial organisms with commercially available microbial identification kit (BBL Crystal ID®). *Philipp. J. Vet. Med.*

Green, M., Bradley, A., Medley, G. and Browne, W., 2007. Cow, farm and management factors during the dry period that determine the rate of clinical mastitis, *J. Dairy Sci.* 90: 3764–76.

Green, M. J., Green, L. E., Medley, G. F., Schukken, Y. H. and Bradley, A. J., 2002. Influence of dry period bacterial intramammary infection on clinical mastitis in dairy cows. *J. Dairy Sci.* 85 (10): 2589–99. PubMed PMID: 12416812.

Gröhn, Y. T., Wilson, D. J., González, R. N., Hertl, J. A., Schulte, H., Bennett, G., Schukken, Y. H., 2004. Effect of pathogen-specific clinical mastitis on milk yield in dairy cows. *J. Dairy Sci.* 87, 3358–74. doi:10.3168/jds.S0022-0302(04)73472-4.

Guélat-Brechbuehl, M., Thomann, A., Albini, S., Moret-Stalder, S., Reist, M., Bodmer, M., Michel, A., Niederberger, M. D. and Kaufmann, T., 2010. Cross-sectional study of Streptococcus species in quarter milk samples of dairy cows in the canton of Bern, Switzerland. *Vet. Rec.* 167, 211–15. doi:10.1136/vr.c4237.

Gurjar, A., Gioia, G., Schukken, Y., Welcome, F., Zadoks, R. and Moroni, P., 2012. Molecular diagnostics applied to mastitis problems on dairy farms. *Vet. Clin. North Am. Food Anim. Pract.* 28, 565–76. doi:10.1016/j.cvfa.2012.07.011.

Haddadi, K., Moussaoui, F., Hebia, I., Laurent, F. and Le Roux, Y., 2005. E. coli proteolytic activity in milk and casein breakdown. *Reprod. Nutr. Dev.* 45, 485–96. doi:10.1051/rnd:2005033.

Halasa, T., Osterås, O., Hogeveen, H., van Werven, T. and Nielen, M., 2009. Meta-analysis of dry cow management for dairy cattle. Part 1. Protection against new intramammary infections. *J. Dairy Sci.* July 92 (7): 3134–49. doi: 10.3168/jds.2008-1740. PubMed PMID: 19528591.

Hamann, J., 2005. *Diagnosis of Mastitis and Indicators of Milk Quality, Mastitis in Dairy Production; Current Knowledge and Future Solutions.* Wageningen Academic Publishers, Wageningen, the Netherlands.

Hamann, J., 1996. Somatic cells: factors of influence and practical measures to keep a physiological level. *Mastit. Newsl.* 21, 9–11.

Harmon, R. J., 1994. Physiology of mastitis and factors affecting somatic cell counts. *J. Dairy Sci.* 77, 2103–12. doi:10.3168/jds.S0022-0302(94)77153-8.

Harrison, E., Bonhotal, J. and Schwarz, M., 2008. Using manure solids as bedding: Final report. Cornell Univ. Waste Manag. Institute, Ithaca, NY.

Hertl, J. A, Gröhn, Y. T., Leach, J. D. G., Bar, D., Bennett, G. J., González, R. N., Rauch, B. J., Welcome, F. L., Tauer, L. W. and Schukken, Y. H., 2010. Effects of clinical mastitis caused by gram-positive and gram-negative bacteria and other organisms on the probability of conception in New York State Holstein dairy cows. *J. Dairy Sci.* 93, 1551–60. doi:10.3168/jds.2009-2599.

Hertl, J. A., Schukken, Y. H., Bar, D., Bennett, G. J., González, R. N., Rauch, B. J., Welcome, F. L., Tauer, L. W. and Gröhn, Y. T., 2011. The effect of recurrent episodes of clinical mastitis caused by gram-positive and gram-negative bacteria and other organisms on mortality and culling in Holstein dairy cows. *J. Dairy Sci.* 94, 4863–77. doi:10.3168/jds.2010-4000.

Hertl, J. A., Schukken, Y. H., Welcome, F. L., Tauer, L. W. and Gröhn, Y. T., 2014. Pathogen-specific effects on milk yield in repeated clinical mastitis episodes in Holstein dairy cows. *J. Dairy Sci.* 97: 1465–80. doi:10.3168/jds.2013-7266.

Hillerton, J. and Berry, E., 2005. Treating mastitis in the cow: a tradition or an archaism?, *J. Appl. Microbiol.* 98: 1250–5.

Hiss, S., Mielenz, M., Bruckmaier, R. M.and Sauerwein, H., 2004. Haptoglobin concentrations in blood and milk after endotoxin challenge and quantification of mammary Hp mRNA expression. *J. Dairy Sci.* 87, 3778–84. doi:10.3168/jds.S0022-0302(04)73516-X.

Hiss, S., Mueller, U., Neu-Zahren, A. and Sauerwein, H., 2007. Haptoglobin and lactate dehydrogenase measurements in milk for the identification of subclinically diseased udder quarters. *Vet. Med.* (Praha). 52, 245–52.

Hogan, J. S. and Smith, K. L., 1997. Occurrence of clinical and subclinical environmental streptoccal mastitis, in: Proceedings of the Symposium on Udder Health Management for Environmental Streptococci. Ontario Veterinary College, Canada, pp. 36–41.

Holdaway, R. J., 1990. A comparison of methods for the diagnosis of bovine subclinical mastitis within New Zealand dairy herds. PhD Thesis, Massey University.

Holdaway, R. J., Holmes, C. W. and Steffert, I. J., 1996. A comparison of indirect methods for diagnosis of subclinical intramammary infection in lactating dairy cows. Part 2: the discriminative ability of eight parameters in foremilk from individual quarters and cows. *Aust. J. Dairy Tech.* 51, 72–8.

Holm, C., Jepsen, L., Larsen, M. and Jespersen, L., 2004. Predominant microflora of downgraded Danish bulk tank milk. *J. Dairy Sci.* 87, 1151–7. doi:10.3168/jds.S0022-0302(04)73263-4.

Hovinen, M., Aisla, A.-M. and Pyörälä, S., 2006. Accuracy and reliability of mastitis detection with electrical conductivity and milk colour measurement in automatic milking. *Acta Agric. Scand. Sect. A – Anim. Sci.* 56, 121–7. doi:10.1080/09064700701216888.

Huerre, M., Ravisse, P., Solomon, H., Ave, P., Briquelet, N., Maurin, S. and Wuscher, N., 1993. [Human protothecosis and environment]. *Bull. la Société Pathol. Exot.* 86, 484–8.

Husfeldt, A. W., Endres, M. I., Salfer, J. A. and Janni, K. A., 2012. Management and characteristics of recycled manure solids used for bedding in Midwest freestall dairy herds. *J. Dairy Sci.* 95, 2195–203. doi:10.3168/jds.2011-5105.

Ibeagha-awemu, E. M., Ibeagha, A. E., Messier, S. and Zhao, X., 2010. Proteomics, genomics, and pathway analyses of *Escherichia coli* and Staphylococcus aureus infected milk whey reveal molecular pathways and networks involved in mastitis. *J. Proteome Res.* 4604–19. doi:10.1021/pr100336e.

Jagielski, T., Lassa, H., Ahrholdt, J., Malinowski, E. and Roesler, U., 2011. Genotyping of bovine Prototheca mastitis isolates from Poland. *Vet. Microbiol.* 149, 283–7. doi:10.1016/j.vetmic.2010.09.034.

Jánosi, S., Rátz, F., Szigeti, G., Kulcsár, M., Kerényi, J., Laukó, T., Katona, F. and Huszenicza, G., 2001. Review of the microbiological, pathological, and clinical aspects of bovine mastitis caused by the alga Prototheca zopfii. *Vet. Q.* 23, 58–61. doi:10.1080/01652176.2001.9695082.

Jayarao, B. M., Oliver, S. P., Tagg, J. R. and Matthews, K. R., 1991. Genotypic and phenotypic analysis of Streptococcus uberis isolated from bovine mammary secretions. *Epidemiol. Infect.* 107, 543–55.

Karlsson, A. and Arvidson, S., 2002. Variation in extracellular protease production among clinical isolates of Staphylococcus aureus due to different levels of expression of the protease repressor sarA. *Infect. Immun.* 70, 4239–46.

Katholm, J. and Andersen, P. H., 1992. Acute coliform mastitis in dairy cows: endotoxin and biochemical changes in plasma and colony-forming units in milk. *Vet. Rec.* 131, 513–14.

Kishimoto, Y., Kano, R., Maruyama, H., Onozaki, M., Makimura, K., Ito, T., Matsubara, K., Hasegawa, A. and Kamata, H., 2010. 26S rDNA-based phylogenetic investigation of Japanese cattle-associated Prototheca zopfii isolates. *J. Vet. Med. Sci.* 72, 123–6.

Kitchen, B. J., 1981. Review of the progress of dairy science: bovine mastitis: milk compositional changes and related diagnostic tests. *J. Dairy Res.* 48, 167–88.

Kossaibati, M. A. and Esslemont, R. J., 1997. The costs of production diseases in dairy herds in England. *Vet. J.* 154, 41–51.

Kuang, Y., Tani, K., Synnott, A. J., Ohshima, K., Higuchi, H., Nagahata, H. and Tanji, Y., 2009. Characterization of bacterial population of raw milk from bovine mastitis by culture-independent PCR–DGGE method. *Biochem. Eng. J.* 45, 76–81. doi:10.1016/j.bej.2009.02.010.

Lacetera, N., Scalia, D., Bernabucci, U., Ronchi, B., Pirazzi, D. and Nardone, A., 2005. Lymphocyte functions in overconditioned cows around parturition. *J. Dairy Sci.* 88 (6): 2010–16. PubMed PMID: 15905431.

Laevens, H., Deluyker, H., Schukken, Y. H., De Meulemeester, L., Vandermeersch, R., De Muêlenaere, E. and De Kruif, A., 1997. Influence of parity and stage of lactation on the somatic cell count in bacteriologically negative dairy cows. *J. Dairy Sci.* 80, 3219–26. doi:10.3168/jds.S0022-0302(97)76295-7.

Lago, A., Godden, S. M., Bey, R., Ruegg, P. L. and Leslie, K., 2011. The selective treatment of clinical mastitis based on on-farm culture results: II. Effects on lactation performance, including clinical mastitis recurrence, somatic cell count, milk production, and cow survival. *J. Dairy Sci.* 94 (9): 4457–67. doi: 10.3168/jds.2010-4047. PubMed PMID: 21854918.

Lago, A., Godden, S. M., Bey, R., Ruegg, P. L. and Leslie, K., 2011. The selective treatment of clinical mastitis based on on-farm culture results: I. Effects on antibiotic use, milk withholding time, and short-term clinical and bacteriological outcomes. *J. Dairy Sci.* 94 (9): 4441–56. doi: 10.3168/jds.2010-4046. PubMed PMID: 21854917.

Larson, M. A., Weber, A., Weber, A. T. and McDonald, T. L., 2005. Differential expression and secretion of bovine serum amyloid A3 (SAA3) by mammary epithelial cells stimulated with prolactin or lipopolysaccharide. *Vet. Immunol. Immunopathol.* 107, 255–64. doi:10.1016/j.vetimm.2005.05.006.

Lass-Flörl, C. and Mayr, A., 2007. Human protothecosis. *Clin. Microbiol. Rev.* 20, 230–242. doi:10.1128/CMR.00032-06.

Le Maréchal, C., Thiéry, R., Vautor, E. and Le Loir, Y., 2011. Mastitis impact on technological properties of milk and quality of milk products – a review. *Dairy Sci. Technol.* 91, 247–82. doi:10.1007/s13594-011-0009-6.

Lopez-Benavides, M. G., Dohoo, I., Scholl, D., Middleton, J. R. and Perez, R., 2012. Interpreting bacteriological culture results to diagnose bovine intramammary infections. *Natl. Mastit. Counc. Res. Comm. Rep.*

Lopez-Benavides, M. G., Williamson, J. H., Pullinger, G. D., Lacy-Hulbert, S. J., Cursons, R. T. and Leigh, J. A., 2007. Field observations on the variation of Streptococcus uberis populations in a pasture-based dairy farm. *J. Dairy Sci.* 90, 5558–66. doi:10.3168/jds.2007-0194.

Malek dos Reis, C. B., Barreiro, J. R., Mestieri, L., Porcionato, M. A. de F. and dos Santos, M. V., 2013. Effect of somatic cell count and mastitis pathogens on milk composition in Gyr cows. *BMC Vet. Res.* 9, 67. doi:10.1186/1746-6148-9-67.

Malinowski, E., Kłossowska, A., Kaczmarowski, M., Kuźma, K.and Markiewicz, H., 2003. Field trials on the prophylaxis of intramammary infections in pregnant heifers. *Pol. J. Vet. Sci.* 6, 117–24.

Marques, S., Silva, E., Kraft, C., Carvalheira, J., Videira, A., Huss, V. A. R. and Thompson, G., 2008. Bovine mastitis associated with Prototheca blaschkeae. *J. Clin. Microbiol.* 46, 1941–5. doi:10.1128/JCM.00323-08.

Mattila, T., Pyörälä, S. and Sandholm, M., 1986. Comparison of milk antitrypsin, albumin, n-acetyl-beta-D-glucosaminidase, somatic cells and bacteriological analysis as indicators of bovine subclinical mastitis. *Vet. Res. Commun.* 10, 113–24.

MacDonald, K. A. R., 2011. Validation of on-farm mastitis pathogen identification systems and determination of the utility of a decision model to target therapy of clinical mastitis during lactation. PhD Thesis. University of Prince Edward Island.

McDonald, T. L., Larson, M. A., Mack, D. R. and Weber, A., 2001. Elevated extrahepatic expression and secretion of mammary-associated serum amyloid A 3 (M-SAA3) into colostrum. *Vet. Immunol. Immunopathol.* 83, 203–11.

Merin, U., Fleminger, G., Komanovsky, J., Silanikove, N., Bernstein, S. and Leitner, G., 2008. Subclinical udder infection with *Streptococcus dysgalactiae* impairs milk coagulation properties: The emerging role of proteose peptones. *Dairy Sci. Technol.* 88, 407–19. doi:10.1051/dst:2008022.

Middleton, J., Timms, L., Bader, G., Lakritz, G., Luby, C. and Steevens, B., 2005. Effect of prepartum intramammary treatment with pirlimycin hydrochloride on prevalence of early first-lactation mastitis in dairy heifers, *J. Am. Vet. Med. Assoc.* 227, 1969–74.

Miglio, A., Moscati, L., Fruganti, G., Pela, M., Scoccia, E., Valiani, A. and Maresca, C., 2013. Use of milk amyloid A in the diagnosis of subclinical mastitis in dairy ewes. *J. Dairy Res.* 80, 496–502. doi:10.1017/S0022029913000484.

Mitchell, G. E., Rogers, S. A., Houlihan, D. B., Tucker, V. C. and Kitchen, B. J., 1986. The relationship between somatic cell count, composition and manufacturing properties of bulk milk. 1. Composition of farm bulk milk. *Aust. J. Dairy Technol.* 41, 9–12.

Möller, A., Truyen, U. and Roesler, U., 2007. Prototheca zopfii genotype 2: the causative agent of bovine protothecal mastitis? *Vet. Microbiol.* 120, 370–4. doi:10.1016/j.vetmic.2006.10.039.

Mottram, T., Rudnitskaya, A., Legin, A., Fitzpatrick, J. L. and Eckersall, P. D., 2007. Evaluation of a novel chemical sensor system to detect clinical mastitis in bovine milk. *Biosens. Bioelectron.* 22, 2689–93. doi:10.1016/j.bios.2006.11.006.

Munoz, M. A., Ahlstro, M. C., Rauch, B. J. and Zadoks, R. N., 2006. Fecal Shedding of *Klebsiella pneumoniae* by Dairy Cows. *J. Dairy Sci.* 89: 3425–30. DOI: 10.3168/jds.S0022-0302(06)72379-7.

Murakami, M., Dorschner, R. A., Stern, L. J., Lin, K. H. and Gallo, R. L., 2005. Expression and secretion of cathelicidin antimicrobial peptides in murine mammary glands and human milk. *Pediatr. Res.* 57, 10–15. doi:10.1203/01.PDR.0000148068.32201.50.

Nguyen, D. A. and Neville, M. C., 1998. Tight junction regulation in the mammary gland. *J. Mammary Gland Biol. Neoplasia* 3, 233–46.

Nielsen, N. I., Larsen, T., Bjerring, M. and Ingvartsen, K. L., 2005. Quarter health, milking interval, and sampling time during milking affect the concentration of milk constituents. *J. Dairy Sci.* 88, 3186–200. doi:10.3168/jds.S0022-0302(05)73002-2.

NMC, 2008. Summary of peer-reviewed publications on efficacy of premilking and postmilking teat disinfectants published since 1980, National Mastitis Council (http://www.nmconline.org/docs/teatbibl.pdf)

Nomura, M., Kobayashi, M., Narita, T., Kimoto-Nira, H. and Okamoto, T., 2006. Phenotypic and molecular characterization of Lactococcus lactis from milk and plants. *J. Appl. Microbiol.* 101, 396–405. doi:10.1111/j.1365-2672.2006.02949.x.

Noordhuizen, J. and Jorritsma, R., 2005. The role of animal hygiene and animal health in dairy operations, International Society for Animal Hygiene Conference, Warsaw, Poland.

Norberg, E., 2005. Electrical conductivity of milk as a phenotypic and genetic indicator of bovine mastitis: a review. *Livest. Prod. Sci.* 96, 129–39. doi:10.1016/j.livprodsci.2004.12.014.

O'Mahony, M. C., Healy, A. M., Harte, D., Walshe, K. G., Torgerson, P. R. and Doherty, M. L., 2006. Milk amyloid A: correlation with cellular indices of mammary inflammation in cows with normal and raised serum amyloid A. *Res. Vet. Sci.* 80, 155–61. doi:10.1016/j.rvsc.2005.05.005.

Odierno, L., Calvinho, L., Traverssa, P., Lasagno, M., Bogni, C. and Reinoso, E., 2006. Conventional identification of Streptococcus uberis isolated from bovine mastitis in Argentinean dairy herds. *J. Dairy Sci.* 89, 3886–90. doi:10.3168/jds.S0022-0302(06)72431-6.

Oliveira, C. A. F., Fernandes, A. M., Neto, O. C. C., Fonseca, L. F. L., Silva, E. O. T. and Balian, S. C., 2002. Composition and sensory evaluation of whole yogurt produced from milk with different somatic cell counts. *Aust. J. Dairy Technol.* 57, 192–6.

Oliveira, L. and Ruegg, P., 2014. Treatments of clinical mastitis occurring in cows on 51 large dairy herds in Wisconsin, *J. Dairy Sci.* 97: 5426–36.

Oliver, S. P. and Murinda, S. E., 2012. Antimicrobial resistance of mastitis pathogens. *Vet. Clin. North Am. Food Anim. Pract.* 28, 165–85. doi:10.1016/j.cvfa.2012.03.005.

Oliver, S. P., Lewis, M. J., Gillespie, B. E. and Dowlen, H. H., 1992. Influence of prepartum antibiotic therapy on intramammary infections in primigravid heifers during early lactation. *J Dairy Sci.* February 75 (2): 406–14. PubMed PMID: 1560135.

Pemberton, R. M., Hart, J. P. and Mottram, T. T., 2001. An assay for the enzyme N-acetyl-beta-D-glucosaminidase (NAGase) based on electrochemical detection using screen-printed carbon electrodes (SPCEs). *Analyst* 126, 1866–71.

Pieper, L., Godkin, A., Roesler, U., Polleichtner, A., Slavic, D., Leslie, K. E. and Kelton, D. F., 2012. Herd characteristics and cow-level factors associated with Prototheca mastitis on dairy farms in Ontario, Canada. *J. Dairy Sci.* 95, 5635–44. doi:10.3168/jds.2011-5106.

Piepers, S., Schukken, Y. H., Passchyn, P. and De Vliegher, S., 2013. The effect of intramammary infection with coagulase-negative staphylococci in early lactating heifers on milk yield throughout first lactation revisited. *J. Dairy Sci.* 96, 5095–105. doi:10.3168/jds.2013-6644.

Piessens, V., Van Coillie, E., Verbist, B., Supré, K., Braem, G., Van Nuffel, A., De Vuyst, L., Heyndrickx, M. and De Vliegher, S., 2011. Distribution of coagulase-negative Staphylococcus species from milk and environment of dairy cows differs between herds. *J. Dairy Sci.* 94, 2933–44. doi:10.3168/jds.2010-3956.

Plumed-ferrer, C., Gazzola, S., Fontana, C., Bassi, D., Cocconcelli, P. and Wright, V., 2015. Strain Isolated from Bovine Mastitis 3, 5–6. doi:10.1128/genomeA.00449-15.Copyright.

Plumed-Ferrer, C., Uusikylä, K., Korhonen, J. and von Wright, A., 2013. Characterization of Lactococcus lactis isolates from bovine mastitis. *Vet. Microbiol.* 167, 592–9. doi:10.1016/j.vetmic.2013.09.011.

Pot, B., Devriese, L. A., Ursi, D., Vandamme, P., Haesebrouck, F. and Kersters, K., 1996. Phenotypic identification and differentiation of Lactococcus strains isolated from animals. *Syst. Appl. Microbiol.* 19, 213–22. doi:10.1016/S0723-2020(96)80047-6.

Pryor, S. M., Cursons, R. T., Williamson, J. H. and Lacy-Hulbert, S. J., 2009. Experimentally induced intramammary infection with multiple strains of Streptococcus uberis. *J. Dairy Sci.* 92, 5467–75. doi:10.3168/jds.2009-2223.

Pyorala, S., 2003. Indicators of inflammation in the diagnosis of mastitis. *Vet. Res.* 34, 565–78.

Pyrola, S., 2009. Treatment of mastitis during lactation, *Ir. Vet. J.* 62 (Suppl. 4): S40–4.

Reinhardt, T. A., Sacco, R. E., Nonnecke, B. J. and Lippolis, J. D., 2013. Bovine milk proteome: quantitative changes in normal milk exosomes, milk fat globule membranes and whey proteomes resulting from Staphylococcus aureus mastitis. *J. Proteomics* 82, 141–54. doi:10.1016/j.jprot.2013.02.013.

Rhoda, D. A. and Pantoja, J. C. F., 2012. Using mastitis records and somatic cell count data. *Vet. Clin. North Am. Food Anim. Pract.* 28, 347–61. doi:10.1016/j.cvfa.2012.03.012.

Ricchi, M., De Cicco, C., Buzzini, P., Cammi, G., Arrigoni, N., Cammi, M. and Garbarino, C., 2013. First outbreak of bovine mastitis caused by Prototheca blaschkeae. *Vet. Microbiol.* 162, 997–9. doi:10.1016/j.vetmic.2012.11.003.

Ricchi, M., Goretti, M., Branda, E., Cammi, G., Garbarino, C. A., Turchetti, B., Moroni, P., Arrigoni, N. and Buzzini, P., 2010. Molecular characterization of Prototheca strains isolated from Italian dairy herds. *J. Dairy Sci.* 93, 4625–31. doi:10.3168/jds.2010-3178.

Roberson, J. R., Warnick, L. D. and Moore, G., 2004. Mild to moderate clinical mastitis: efficacy of intramammary amoxicillin, frequent milk-out, a combined intramammary amoxicillin, and frequent milk-out treatment versus no treatment. *J. Dairy Sci.* 87, 583–592. doi:10.3168/jds.S0022-0302(04)73200-2.

Roesler, U. and Hensel, A., 2003. Longitudinal analysis of Prototheca zopfii-specific immune responses: correlation with disease progression and carriage in dairy cows. *J. Clin. Microbiol.* 41, 1181–6.

Roesler, U., Scholz, H. and Hensel, A., 2001. Immunodiagnostic identification of dairy cows infected with Prototheca zopfii at various clinical stages and discrimination between infected and uninfected cows. *J. Clin. Microbiol.* 39, 539–43. doi:10.1128/JCM.39.2.539-543.2001.

Rogers, S. A., Mitchell, G. E. and Bartley, J. P., 1994. The relationship between somatic cell count, composition and manufacturing properties of bulk milk 6. Cheddar cheese and skim milk yoghurt. *Aust. J. Dairy Technol.* 49, 70–4.

Ruegg, P. L., 2003. Investigation of mastitis problems on farms. *Vet. Clin. North Am. Food Anim. Pract.* 19, 47–73.

Ruegg, P. L., 2011. Managing mastitis and producing quality milk. *Dairy Prod. Med.* 207–32.

Ruegg, P. L. and Pantoja, J. C. F., 2013. Understanding and using somatic cell counts to improve milk quality. *Irish J. Agric. Food Res.* 52, 101–17.

Santos, J. E. P., Juchem, S. O., Cerri, R. L. A., Galvão, K. N., Chebel, R. C., Thatcher, W. W., Dei, C. S. and Bilby, C. R., 2004. Effect of bST and reproductive management on reproductive performance of Holstein dairy cows. *J. Dairy Sci.* 87, 868–81. doi:10.3168/jds.S0022-0302(04)73231-2.

Scaccabarozzi, L., Locatelli, C., Pisoni, G., Manarolla, G., Casula, A., Bronzo, V. and Moroni, P., 2011. Short communication: Epidemiology and genotyping of Candida rugosa strains responsible for persistent intramammary infections in dairy cows. *J. Dairy Sci.* 94, 4574–7. doi:10.3168/jds.2011-4294.

Schepers, A. J., Lam, T. J., Schukken, Y. H., Wilmink, J. B. and Hanekamp, W. J., 1997. Estimation of variance components for somatic cell counts to determine thresholds for uninfected quarters. *J. Dairy Sci.* 80, 1833–40. doi:10.3168/jds.S0022-0302(97)76118-6.

Scherpenzeel, C. G., Tijs, S. H., den Uijl, I. E., Santman-Berends, I. M., Velthuis, A. G. and Lam, T. J., 2016. Farmers' attitude toward the introduction of selective dry cow therapy. *J Dairy Sci.* 99 (10): 8259–66. doi: 10.3168/jds.2016-11349. PubMed PMID: 27448856.

Schukken, Y. H., Bennett, G. J., Zurakowski, M. J., Sharkey, H. L., Rauch, B. J., Thomas, M. J., Ceglowski, B., Saltman, R. L., Belomestnykh, N. and Zadoks, R. N., 2011. Randomized clinical trial to evaluate the efficacy of a 5-day ceftiofur hydrochloride intramammary treatment on nonsevere gram-negative clinical mastitis. *J. Dairy Sci.* 94, 6203–15. doi:10.3168/jds.2011-4290.

Schukken, Y. H., Hertl, J., Bar, D., Bennett, G. J., González, R. N., Rauch, B. J., Santisteban, C., Schulte, H. F., Tauer, L., Welcome, F. L. and Gröhn, Y. T., 2009. Effects of repeated gram-positive and gram-negative clinical mastitis episodes on milk yield loss in Holstein dairy cows. *J. Dairy Sci.* 92, 3091–105. doi:10.3168/jds.2008-1557.

Schukken, Y. H., Günther, J., Fitzpatrick, J., Fontaine, M. C., Goetze, L., Holst, O., Leigh, J., Petzl, W., Schuberth, H. J., Sipka, A., Smith, D. G., Quesnell, R., Watts, J., Yancey, R., Zerbe, H., Gurjar, A., Zadoks, R. N. and Seyfert, H. M., 2011. members of the Pfizer mastitis research consortium.. Host-response patterns of intramammary infections in dairy cows. *Vet Immunol. Immunopathol.* December 15; 144 (3–4): 270–89. doi: 10.1016/j.vetimm.2011.08.022. Review. PubMed PMID: 21955443.

Schukken, Y. H., Wilson, D. J., Welcome, F., Garrison-Tikofsky, L. and Gonzalez, R. N., 2003. Monitoring udder health and milk quality using somatic cell counts. *Vet. Res.* 34, 579–96. doi:10.1051/vetres:2003028.

Sharma, K. K. and Randolph, H. E., 1974. Influence of mastitis on properties of milk. 8. Distribution of soluble and micellar casein. *J. Dairy Sci.* 57, 19–23.

Shuster, D. E., Harmon, R. J., Jackson, J. A. and Hemken, R. W., 1991. Suppression of milk production during endotoxin-induced mastitis. *J. Dairy Sci.* 74, 3763–74. doi:10.3168/jds.S0022-0302(91)78568-8.

Small, A., 2006. Hygiene of milk and dairy products, in Buncic, S. (ed.), *Integrated Food Safety and Veterinary Public Health*, CABI Publishers, Wallingford, UK.

Smith, K. L., Todhunter, D. A. and Schoenberger, P. S., 1985. Environmental pathogens and intramammary infection during the dry period. *J. Dairy Sci.* 68 (2): 402–17. PubMed PMID: 4039338.

Smith, A., Westgarth, D. R., Jones, M. R., Neave, F. K., Dodd, F. H. and Brander, G. C., 1967. Methods of reducing the incidence of udder infection in dry cows. *Vet Rec.* November 11; 81 (20): 504–10. PubMed PMID: 5624737.

Smolenski, G. A., Wieliczko, R. J., Pryor, S. M., Broadhurst, M. K., Wheeler, T. T. and Haigh, B. J., 2011. The abundance of milk cathelicidin proteins during bovine mastitis. *Vet. Immunol. Immunopathol.* 143, 125–30. doi:10.1016/j.vetimm.2011.06.034.

Smolenski, G., Haines, S., Kwan, F. Y. S., Bond, J., Farr, V., Davis, S. R., Stelwagen, K. and Wheeler, T. T., 2007. Characterisation of host defense proteins in milk using a proteomic approach. *J. Proteome Res.* 6, 207–15.

Smolenski, G. A., Broadhurst, M. K., Stelwagen, K., Haigh, B. J. and Wheeler, T. T., 2014. Host defence related responses in bovine milk during an experimentally induced Streptococcus uberis infection. *Proteome Sci.* 12, 19. doi:10.1186/1477-5956-12-19.

Sordillo, L. M., Contreras, G. A. and Aitken, S. L., 2009. Metabolic factors affecting the inflammatory response of periparturient dairy cows. *Anim. Health Res. Rev.* June; 10 (1): 53–63. doi: 10.1017/S1466252309990016. Review. PubMed PMID: 19558749.

Spanamberg, A., Wünder, E. A., Brayer Pereira, D. I., Argenta, J., Cavallini Sanches, E. M., Valente, P. and Ferreiro, L., 2008. Diversity of yeasts from bovine mastitis in Southern Brazil. Rev. *Iberoam. Micol.* 25, 154–6.

Suojala, L., Kaartinen, L. and Pyorala, S., 2013. Treatment for bovine *Escherichia coli* mastitis – an evidence-based approach, *Vet. Pharm. Ther.* 36, 521–31.

Supré, K., Haesebrouck, F., Zadoks, R. N., Vaneechoutte, M., Piepers, S. and De Vliegher, S., 2011. Some coagulase-negative Staphylococcus species affect udder health more than others. *J. Dairy Sci.* 94, 2329–40. doi:10.3168/jds.2010-3741.

Svec, P. and Sedlácek, I., 2008. Characterization of Lactococcus lactis subsp. lactis isolated from surface waters. *Folia Microbiol.* (Praha). 53, 53–6. doi:10.1007/s12223-008-0007-0.

Teixeira, L. M., Merquior, V. L., Vianni, M. C., Carvalho, M. G., Fracalanzza, S. E., Steigerwalt, A. G., Brenner, D. J. and Facklam, R. R., 1996. Phenotypic and genotypic characterization of atypical Lactococcus garvieae strains isolated from water buffalos with subclinical mastitis and confirmation of L. garvieae as a senior subjective synonym of Enterococcus seriolicida. *Int. J. Syst. Bacteriol.* 46, 664–8. doi:10.1099/00207713-46-3-664.

Tenhagen, B., Kalbe, P., Baumgartner, B. and Heuwieser, W., 2001. An outbreak of mastitis caused by Prototheca zopfii on a large confinement dairy: analysis of cow level risk factors. Proc. 2nd Int. Symp. Mastit. Milk Qual.

Thompson, G., Silva, E., Marques, S., Müller, A. and Carvalheira, J., 2009. Algaemia in a dairy cow by Prototheca blaschkeae. *Med. Mycol.* 47, 527–31. doi:10.1080/13693780802566341.

Timms, L., 2000. Field trial evaluations of a persistent barrier teat dip for preventing mastitis during the dry period, in *Proc. Symp. Immunol. Ruminant Mammary Gland*, Stresa, Italy.

Todhunter, D. A., Smith, K. L. and Hogan, J. S., 1995. Environmental streptococcal intramammary infections of the bovine mammary gland. *J. Dairy Sci.* 78, 2366–74. doi:10.3168/jds.S0022-0302(95)76864-3.

Trevisi, E., Amadori, M., Cogrossi, S., Razzuoli, E. and Bertoni, G., 2012. Metabolic stress and inflammatory response in high-yielding, periparturient dairy cows. *Res. Vet Sci.* 93 (2): 695–704. doi: 10.1016/j.rvsc.2011.11.008. PubMed PMID: 22197526.

Vanderhaeghen, W., Piepers, S., Leroy, F., Van Coillie, E., Haesebrouck, F. and De Vliegher, S., 2015. Identification, typing, ecology and epidemiology of coagulase negative staphylococci associated with ruminants. *Vet. J.* 203, 44–51. doi:10.1016/j.tvjl.2014.11.001.

Vanderhaeghen, W., Piepers, S., Leroy, F., Van Coillie, E., Haesebrouck, F. and De Vliegher, S., 2014. Invited review: effect, persistence, and virulence of coagulase-negative Staphylococcus species associated with ruminant udder health. *J. Dairy Sci.* 97, 5275–93. doi:10.3168/jds.2013-7775.

Vasquez, A. K., Nydam, D. V., Capel, M. B., Ceglowski, B., Rauch, B. J., Thomas, M. J., Tikofsky, L., Watters, R. D., Zuidhof, S. and Zurakowski, M. J., 2016. Randomized noninferiority trial comparing 2 commercial intramammary antibiotics for the treatment of nonsevere clinical mastitis in dairy cows. *J. Dairy Sci.* 99 (10): 8267–81.

Viguier, C., Arora, S., Gilmartin, N., Welbeck, K. and O'Kennedy, R., 2009. Mastitis detection: current trends and future perspectives. *Trends Biotechnol.* 27, 486–93. doi:10.1016/j.tibtech.2009.05.004.

Villarroel, A., Dargatz, D., Lane, V., McCluskey, B. and Salman, M., 2007. Suggested outline of potential critical control points for biosecurity and biocontainment on large dairy farms, *J. Am. Vet. Med. Assoc.* 230: 808–19.

Vosough Ahmadi, B., Frankena, K., Turner, J., Velthius, A., Hogeveen, H. and Huirne, R., 2007. Effectiveness of simulated interventions in reducing the estimated prevalence of *E. coli* O157 in lactating cows in dairy herds, *Vet. Res.* 38: 755–71.

Wenz, J. R., Barrington, G. M., Garry, F. B., McSweeney, K. D., Dinsmore, R. P., Goodell, G. and Callan, R. J., 2001. Bacteremia associated with naturally occuring acute coliform mastitis in dairy cows. *J. Am. Vet. Med. Assoc.* 219, 976–81.

Wenz, J. R., Garry, F. B., Lombard, J. E., Elia, R., Prentice, D. and Dinsmore, R. P., 2005. Short communication: efficacy of parenteral ceftiofur for treatment of systemically mild clinical mastitis in dairy cattle. *J. Dairy Sci.* 88, 3496–9. doi:10.3168/jds.S0022-0302(05)73034-4.

Werner, B., Moroni, P., Gioia, G., Lavín-Alconero, L., Yousaf, A., Charter, M. E., Carter, B. M., Bennett, J., Nydam, D. V, Welcome, F. and Schukken, Y. H., 2014. Short communication: genotypic and phenotypic identification of environmental streptococci and association of Lactococcus lactis ssp. lactis with intramammary infections among different dairy farms. *J. Dairy Sci.* 97, 6964–9. doi:10.3168/jds.2014-8314.

Wheeler, T. T., Smolenski, G. A., Harris, D. P., Gupta, S. K., Haigh, B. J., Broadhurst, M. K., Molenaar, A. J. and Stelwagen, K., 2012. Host-defence-related proteins in cows' milk. *Animal* 6, 415–22. doi:10.1017/S1751731111002151.

Whist, A. C., Østerås, O. and Sølverød, L., 2006. Clinical mastitis in Norwegian herds after a combined selective dry-cow therapy and teat-dipping trial. *J. Dairy Sci.* 89 (12): 4649–59. PubMed PMID: 17106097.

Whyte, D., Walmsley, M., Liew, A., Claycomb, R. and Mein, G., 2005. Chemical and rheological aspects of gel formation in the California Mastitis Test. *J. Dairy Res.* 72, 115–21.

Wiesner, J. and Vilcinskas, A., 2010. Antimicrobial peptides – The ancient arm of the human immune system. *Virulence* 1, 440–64.

Wiess, W., 2002. 'Relationship of mineral and vitamin supplementation with mastitis and milk quality', National Mastitis Council (http://www.nmconline.org/articles/nutr.pdf)

Wilson, D., Mallard, B., Burton, J., Schukken. Y. and Grohn, Y., 2009. Association of *Escherichia coli* J5-specific serum antibody responses with clinical mastitis outcome for J5 vaccinate and control dairy cattle, *Clin. Vaccine Immunol.* 16(2): 209–17.

Zadoks, R. N. and Fitzpatrick, J. I., 2009. Changing trends in mastitis. *Ir. Vet. J.* 62 Suppl 4, S59–70. doi:10.1186/2046-0481-62-S4-S59.

Zanetti, M., 2005. The role of cathelicidins in the innate host defenses of mammals. *Curr. Issues Mol. Biol.* 7, 179–96.

Zanetti, M., 2004. Cathelicidins, multifunctional peptides of the innate immunity. *J. Leukoc. Biol.* 75, 39–48. doi:10.1189/jlb.0403147.Journal.

Zaragoza, C. S., Olivares, R. A. C., Watty, A. E. D., Moctezuma, A. de la, P. and Tanaca, L. V., 2011. Yeasts isolation from bovine mammary glands under different mastitis status in the Mexican High Plateu. Rev. Iberoam. *Micol.* 28, 79–82. doi:10.1016/j.riam.2011.01.002.

Zhang, L., Boeren, S., van Hooijdonk, A. C. M., Vervoort, J. M. and Hettinga, K. A., 2015. A proteomic perspective on the changes in milk proteins due to high somatic cell count. *J. Dairy Sci.* 98, 5339–51. doi:10.3168/jds.2014-9279.

Zhang, S. and Maddox, C. W., 2000. Cytotoxic activity of coagulase-negative staphylococci in bovine mastitis. *Infect. Immun.* 68, 1102–8.

Chapter 2

Mastitis, milk quality and yield

P. Moroni, Cornell University, USA and University of Milano, Italy; F. Welcome, Cornell University, USA; and M. F. Addis, Porto Conte Ricerche, Italy

1 Introduction

Mastitis is one of the most economically important diseases in dairy production, and it is defined as an inflammation of the mammary gland. Intramammary infections (IMI) continue to be the most important cause of mastitis in dairy cattle, accounting for 38% of the total costs of the common production diseases (Kossaibati and Esslemont, 1997). In the last decade, several groups have estimated the losses associated with clinical mastitis, and the average costs per case (US$) of Gram-positive, Gram-negative and other microorganisms were $133.73, $211.03 and $95.31, respectively (Cha et al., 2013; Gröhn et al., 2004). These costs include treatment, culling, death and decreased milk production. In addition to reduced cow welfare and increased veterinary costs, episodes of mastitis are associated with reduction of milk production (Bar et al., 2007; Schukken et al., 2009), decreased fertility (Hertl et al., 2010; Santos et al., 2004), and increased culling and death risk (Hertl et al., 2011).

Increased somatic cell count (SCC) is considered as a reaction to inflammation (Harmon, 1994) in lactating mammary glands. Normal milk does contain few somatic cells, and the number is almost always lower than 100 000 cells/mL in milk from heifers uninfected/uninflamed mammary quarters or less than 200 000 cells/mL in milk from mature cows (Dohoo and Meek, 1982; Hamann, 1996). Since mastitis is most often due to a bacterial IMI (Djabri et al., 2002), the terms IMI and subclinical mastitis have been used interchangeably (Barkema et al., 1997; Deluyker et al., 2005). According to a recent document by the National Mastitis Council (Lopez-Benavides et al., 2012), the terms 'mastitis' and 'intramammary infection' should not be used interchangeably, as they represent different

http://dx.doi.org/10.19103/AS.2016.0005.20

Table 1 Definition of intramammary infection and mastitis

	Intramammary infection	Mastitis
International Dairy Federation definition	An infection occurring in the secretory tissue and/or the ducts and tubules of the mammary gland	Inflammation of one or more quarters of the mammary gland, almost always caused by infecting microorganisms

Source: Lopez-Benavidez et al., 2012.

entities (Table 1), and the definitions provided by the International Dairy Federation should always be used when referring to these conditions (Lopez-Benavides et al., 2012).

1.1 Clinical mastitis

Clinical mastitis is an inflammatory response to infection causing visibly abnormal milk (e.g. colour, fibrin clots and watery appearance). If clinical cases include only visible changes in the appearance of milk, notable swelling or painful udder, the cases are classified as mild or moderate in severity. If the inflammatory response includes systemic involvement (e.g. fever, anorexia and shock), the case is categorized as severe. Assigning a severity score to individual clinical cases along with identification of the pathogen involved in the case (culture result) are used by veterinarians to assign specific treatment protocols. If the onset is very rapid, as often occurs with severe clinical cases, it is termed an acute case of severe mastitis. More severely affected cows tend to have more serous secretions in the affected quarter. Clinical cases that fall into the severe category account for 10–15% of infections.

Long-term recurring cases of the disease are termed chronic. These cases may show few visible signs of inflammation between repeated occasional clinical flare up of the disease and can continue over periods of several months. Chronic cases of mastitis are often associated with irreversible damage of the udder tissues from the repeated clinical occurrences of the illness. These cows are usually culled.

1.2 Subclinical mastitis

Subclinical mastitis is generally caused by the presence of an infection without any apparent signs of local inflammation or systemic involvement. Even if episodes of abnormal milk or udder inflammation may appear, these infections are generally asymptomatic and, if the infection persisted for at least 2 months, with an increased SCC of milk >200 000 cells/mL are termed chronic. The majority of these infections persist for entire lactations or the life of the cow. However, subclinical mastitis implies inflammation within the udder, but not necessarily infection. Different pathogens are associated with it, especially *Staphylococcus aureus*. Subclinical inflammation is important as cows can continue to shed microorganisms within the rest of the herd, with pathogen spread from cow to cow during milking.

2 Indicators of mastitis

The most established and widely recognized method for mastitis monitoring at the cow and herd level consists in measuring the cells that are present in milk, that is, determining its SCC. The SCC is defined as the number of cells per millilitre of milk (cells/mL) (Dohoo and Leslie, 1991; Ruegg and Pantoja, 2013). A quarter with SCC above 200 000 cells/mL in mature cows (100 000 cells/mL in first lactation cows) is an indication of an inflammatory response and the quarter is likely to be infected. The milk has changed properties such as reduced shelf life of fluid milk and reduced yield and quality of cheese (Barbano et al., 2006).

The SCC can be measured in bulk tank milk (BMSCC), at cow level with composite samples of all four quarters (CSSCC) and at quarter level (QMSCC). BMSCC values are the reference for defining national and international standards for hygienic production of milk. Regulatory standards for comingled milk (BMSCC) may significantly differ depending on the country, ranging from <400 000 cells/mL (such as in the EU, Australia, New Zealand and Canada) (USDA, 2013) to <500 000 cells/mL (Brazil from 2016), and are currently 750 000 in the United States. The BMSCC is used for monitoring cow health at the herd level. The optimal BMSCC is not definitively described, but it is generally considered to be <250 000 cells/mL for milk premium. Some milk buyers offer milk quality premiums based in part on BMSCC to milk producers to encourage lower BMSCC (Ruegg and Pantoja, 2013).

BMSCC can provide reliable indications at the herd level, but measuring CSSSC or QMSCC is necessary for monitoring udder health at the cow level (De Vliegher et al., 2012). This helps to keep subclinical mastitis under control and to obtain more reliable estimates on mastitis prevalence and incidence. The dynamics of SCC values at both herd and cow levels from dairy herd improvement (DHI) programmes (Laevens et al., 1997; Ruegg, 2003) are used in herd management to identify cows that need interventions including culture, treatment, segregation or removal from the herd (Cook et al., 2002; Rhoda and Pantoja, 2012). QMSCC values of 200 000 cells/mL are currently believed to possess a level of specificity sufficient to provide the least diagnostic error in detecting an IMI (Bradley and Green, 2005; Dohoo and Leslie, 1991; Schepers et al., 1997; Schukken et al., 2003), but lower values may be more adequate if a higher sensitivity is desired (Bradley and Green, 2005; Dohoo and Leslie, 1991; Ruegg and Pantoja, 2013; Schepers et al., 1997; Schukken et al., 2003). In general, the following applies: a QMSCC of 100 000 cells/mL or lower indicates absence of mastitis, while a QMSCC of 200 000 cells/mL or higher indicates presence of mastitis, and therefore IMI.

Somatic cells can be enumerated in milk by means of automated cell counting instrumentation, either in the laboratory or at the milking plant. A wide range of devices are available on the market that can meet different throughputs and requirements. In addition, somatic cells can be assessed with cow-side methods such as the California mastitis test (CMT). Due to its qualitative nature, the CMT is highly subjective and dependent on user experience, especially for SCC below 1 000 000 cells/mL, and therefore it has a low sensitivity. It is, however, highly cost-effective and practical for verifying the status of a cow and individual quarters. Interpreting individual quarter data from the CMT allows managers to select from a number of management options including sampling for culture, treatment, segregation of milk, dry off or culling, to best manage individual cows.

Adding to the use of immune cells as indicators of mastitis, other molecules released in milk as a result of an inflammatory process can represent useful, reliable and practical markers. Several enzymes, sugars and salts, are already known to increase in milk during mastitis (Pyorala, 2003), but the advances in biomarker discovery methods based on proteomic techniques (Abd El-Salam, 2014; Ceciliani et al., 2014; Reinhardt et al., 2013; Smolenski et al., 2014) have more recently enabled the identification of other protein and peptide candidates that can form the basis for novel laboratory and field assays. In addition, advancements in immunological assays, both for the laboratory and for the field, have increased sensitivity and specificity of biomarker detection and can represent inexpensive and practical alternatives (Gurjar et al., 2012; Viguier et al., 2009).

Biosensors and immuno-biosensors have been developed for detecting protein markers of mastitis and other, non-protein, mastitis-associated molecules. Biosensors are analytical devices based on an immobilized biological material (e.g. an enzyme, an antibody) that interacts with the molecule to be detected (e.g. a small molecule, a protein) producing a measurable physical, chemical or electrical signal. Adding to detection in the field, these sensors make it possible to implement marker measurement online. The recent, significant increase of robotic milking would represent a powerful way to implement biosensor-based, online mastitis detection strategies. Currently, online tests are available, based on the SCC, milk colour determination or electrical conductivity (EC) (Hovinen et al., 2006; Norberg, 2005). However, these are neither reliable nor sensitive for a conclusive diagnosis (Viguier et al., 2009). The ability to readily monitor more reliable mastitis markers online with a biosensor during milking would represent a powerful opportunity for the earlier and timely detection of mastitis.

Another useful implementation of protein marker measurement consists in the development of rapid, portable 'cow-side' mastitis tests that can enable a more reliable and a less subjective interpretation of results when compared to the CMT or to measuring the EC of milk with handheld metres, which does not seem to represent a reliable alternative (Pyorala, 2003). Also defined as pen-side, point-of-care or rapid diagnostic tests, these are mostly based on antibody-based techniques, including agglutination, enzyme immunoassays and lateral-flow immunochromatography, and take the form of dipsticks or lateral-flow devices (with the appearance of different symbols or lines depending on the result), latex agglutination systems (coagulation if positive) and in-solution systems (change of solution colour). Usually, cow-side tests do not require dedicated instrumentation for carrying out, reading and interpreting test results, and the reaction occurs in a short time. The main advantage of cow-side tests is that the diagnostic information is readily available where it is needed.

3 Impact of mastitis on milk composition

Control and minimization of mastitis are necessary for consistently producing high-quality milk. In fact, mastitis has a negative effect on its physico-chemical properties and on its relative composition (Auldist and Hubble, 1998; Auldist et al., 1995; Le Maréchal et al., 2011). In addition, due to the reduced milk yield and therefore total volume, the concentration of all components changes also in absolute terms (Kitchen, 1981). Although the extent and quality of such alterations depend on disease severity and on the IMI agent (Pyorala, 2003), mastitic milk shows an increase in total proteins and a decrease in caseins (Auldist and Hubble, 1998; Auldist et al., 1995; Holdaway, 1990; Shuster et al., 1991), modifications in the amount and composition of fats, a decrease in lactose and a fluctuation in the amount of the major ions (Table 2) (Petrovski, K. R. and Stefanov, E. 2006. *Milk Composition Changes*. Massey University, USA, http://www.milkproduction.com/Library/Scientific-articles/Animal-health/Milk-composition-changes/). These changes are due to the increase in vascular permeability caused by the inflammation reaction to the

Table 2 Effect of mastitis on milk components

Milk constituent	Effect
Quarter milk yield	–(– –)
Somatic cell count	+++
Dry matter	–
Total protein	?
Total casein	– –
Alpha-casein	– –
Beta-casein	– – –
Kappa-casein	?
Gamma-casein	+
Non-casein N	+
Whey protein	+++
Alpha-lactalbumin	–
Beta-lactoglobulin	– – –
Serum albumin	+
Lactoferrin	+++
Immunoglobulins	+++
Lysozyme	+++
NAGase	+++
Beta-glucuronidase	+++
Plasmin	+++
Lipase	++
Proteose peptones	++
Total fat	?
Free fatty acids	++
Long-chained fatty acids	–
Short-chained fatty acids	+
Lactate	+++
Lactose	–
Sodium	++
Chloride	++
Potassium	–
Calcium	?
Magnesium	?

+, slight increase; ++, moderate increase; +++, pronounced increase; -, slight decrease; --, moderate decrease; ---, pronounced decrease; ?, controversial.

loss in integrity of the mammary epithelium, to damage of the milk-producing cells, to the increase in number and activity of leucocytes recalled in milk by the circulation and to the enzymatic action of microorganisms.

3.1 Protein

The most evident effect on protein composition is a reduction in the total amount of caseins. This is accompanied by an increase in blood serum proteins, including albumin, immunoglobulins, lactoferrin and alpha-macroglobulin, and by an evident decrease in the main milk serum proteins □-lactalbumin and □-lactoglobulin (Auldist and Hubble, 1998). Adding to changes in the amount of total caseins, their relative ratios are also affected; alpha- and beta-casein decrease while kappa-casein slightly increases. Moreover, a decrease in micellar casein is observed, while soluble casein increases (Sharma and Randolph, 1974).

A higher amount of plasmin is also found in mastitic milk, originating from plasminogen in blood serum. Plasmin, normally present in small amounts in milk, is able to catalyse the rapid cleavage of β-casein, leading to its reduction and to an increase in polypeptide fragments (defined as proteose peptone). This sums up to the effect of proteinases produced by activated immune cells that can act on milk proteins in various ways. Several mastitis-causing bacteria, including *Escherichia coli*, *Staphylococcus aureus* and CNS, do also produce casein-degrading proteases (Devriese et al., 1985; Haddadi et al., 2005; Karlsson and Arvidson, 2002; Zhang and Maddox, 2000).

3.2 Fat

High SCC is also associated to changes in milk fat, but this has been studied less extensively than for proteins, with results are somehow contradictory. In fact, many authors do report an increase in total milk fat (Hiss et al., 2004; Holdaway, 1990; Pyorala, 2003; Shuster et al., 1991), while others report a decrease (Auldist and Hubble, 1998). According to some authors, the increase in fat concentration is mainly due to reduced milk volume, while fat synthesis is not significantly affected (Bruckmaier et al., 2004; Holdaway, 1990). On the other hand, others point out that a reduced synthetic and secretory capacity is expected due to inflammatory damage to milk-producing cells (Auldist and Hubble, 1998). This notwithstanding, quality seems to be affected, as milk fat globule membranes can be attacked by lipases produced by the elevated number of immune cells recalled in milk as a result of the inflammatory response. Therefore, lipid breakdown and oxidation phenomena occur with the development of changes in sensory properties and resulting off-flavours (Auldist and Hubble, 1998; Auldist et al., 1995; Holdaway, 1990).

It is likely that changes in fats are more influenced by the specific causative agent and by disease severity when compared to proteins. Therefore, the conflicting results seen in the literature may derive by the fact that SCC is the only marker considered for defining mastitic milk, and the specific pathogen causing IMI is not taken into account. Nevertheless, according to Malek dos Reis and co-workers (2013), IMI caused by CNS, *Streptococcus* spp. and *Corynebacterium* spp. led to a significant reduction in milk fat content, while Rogers et al. (1994) reported no effect. Auldist et al. (1995) and Kitchen (1981) observed a decrease in fats upon subclinical mastitis, whereas Mitchell et al. (1986) reported an increase.

3.3 Lactose

It is commonly accepted that mastitis results in a reduction of milk lactose (Auldist and Hubble, 1998; Auldist et al., 1995; Bruckmaier et al., 2004; Pyorala, 2003; Shuster et al., 1991), most likely due to the impairment of alveolar epithelial cells and the leakage of lactose out of milk and into the circulation. In fact, the drop in lactose concentration that occurs upon mastitis is believed to be mainly due to damage to the tight junctions and therefore to its diffusion via paracellular pathways (Auldist et al., 1995). To support this hypothesis, elevated concentrations of lactose are found in blood and urine of mastitic cows as lactose is synthesized only by epithelial cells in the mammary alveolus. This provides a measure of the mammary epithelial damage and of the extent of leakage from the mammary gland lumen (Nguyen and Neville, 1998). Another contributor to the decrease of lactose in milk is the ability of many mastitis-causing bacteria to degrade this sugar and therefore contribute to reducing its concentration (Auldist et al., 1995). Accordingly, lactate increases in mastitic milk.

3.4 Ions

The increase in permeability of the mammary epithelium and the reduced milk volume do also contribute to changes in many milk ions, and this has a significant effect on the manufacturing quality of milk. The change in ion composition has been exploited for monitoring mastitis, as described above. One of the main consequences is a change in the potassium to sodium ratio; specifically, potassium decreases at the advantage of sodium, as a result of impairment of the Na/K ATPase and of the increase in epithelial permeability. In fact, the functionality of the pump is compromised reducing potassium influx, while sodium leaks into milk from the bloodstream, where its concentration is higher (Allen, 1990; Auldist and Hubble, 1998; Auldist et al., 1995; Forsbäck et al., 2010). Chloride does also increase in mastitis, due to the influx from blood.

4 Impact of mastitis on dairy product quality

Milk with good quality is needed also for producing good-quality dairy products. In fact, adding to the risk of bacterial contamination, changes in milk composition can have a significant impact on transformation processes, both in terms of yield, physical and sensory properties. Actually, the quality of dairy products can be significantly affected even when the composition of milk is only slightly impaired (Merin et al., 2008).

Deterioration of sensory properties can already occur in pasteurized or UHT milk (Le Maréchal et al., 2011). In fact, heat-treated high SCC milk can present several off-flavours, including rancidity and bitterness, that are mainly due to endogenous alterations in enzymes favouring proteolysis and lipolysis phenomena, as well to the growth of bacteria in milk (Barbano et al., 2006).

According to several authors, presence of an increased SCC has a reduced impact on cow milk yoghurt, and it does have little effect on its acidity, fat and protein composition, or microbiological features (Fernandes et al., 2007; Oliveira et al., 2002), although high SCC can have unfavourable consequences on product shelf life. A loss of consistency and impaired taste were seen after 20 and 30 days at 5°C, respectively (Oliveira et al., 2002). An increase in viscosity during storage was also reported for yoghurts made with high SCC milk (Fernandes et al., 2007).

The significant reduction in total casein at the advantage of other proteins has important and obvious consequences on cheese quality and yield, but other detrimental effects are also seen (Auldist and Hubble, 1998; Auldist et al., 1995). Adding to changes in the amount of total caseins, their relative ratios are also affected; alpha- and beta-casein decrease while kappa-casein does slightly increase, and the main decrease is at the expense of micellar casein while soluble casein increases. Due to these changes, numerous technological aspects can be significantly impacted, although with different extents in different types of cheeses. In general, these include coagulation properties with a significant increase in clotting time, lower curd firmness and slower rate of curd firming; changes in the final moisture content of cheese; and development of various off-flavours. Presence of host- and pathogen-produced proteases in high SCC milk also lead to poor curding, reduced cheese yield and to a series of negative sensory changes including texture, flavour and functionality (Le Maréchal et al., 2011).

Due to the impact of high SCC on renneting time, the yield of some cheeses is reduced (e.g. cottage cheese and cheddar), while others do not seem to be affected (e.g. mozzarella and zamorano). Nevertheless, cheese composition is always affected with higher levels of proteolysis and lypolysis, as well as flavour, body and texture of the final product. For example, in cheddar cheese production, the use of high SCC milk leads to the development of a 'lipolytic' or 'oxidized' flavour (Auldist et al., 1996). Presence of bacteria can also impact cheese ripening by interfering with the correct development of the microbial flora, and therefore to the final sensorial properties (e.g. texture, flavour and odour) of ripened cheese.

5 Impact of mastitis on milk production yield

The precise relationship existing among SCC, immune response, mastitis, IMI and milk production is not so straightforward. In fact, there is a debate concerning slight increases in SCC at quarter milk level due to the presence of minor pathogens *Corynebacterium bovis* (Kurek, 1980) and, more recently, CNS (Piepers et al., 2013).

Kurek has demonstrated that *C. bovis* predisposes the udder to spread infection during the dry period. Herds with low BMSCC have higher incidence of environmental mastitis compared to those with elevated BMSCC (Schukken et al., 1990; Green et al., 1996, 2001; Waage et al., 1998). In support of this hypothesis, moderate individual cow milk SCC protects against experimental infection by environmental mastitis pathogens (Matthews and Harmon, 1989).

SCCs increase with the severity of udder inflammation. However, the corresponding milk loss does not increase at the same rate. In the 1980s, the use of a log-linear score to report SCCs and predict milk loss associated with SCC to dairy producers was proposed. The use of linear scores (LS), also referred to as somatic cell score (SCS) simplifies the prediction of milk loss. The formula for calculating the SCS is log2 (SCC/100 000)+3, where SCC is in units of cells/mL. The resulting score was initially referred to as a linear score owing to the linear relationship between score and milk loss associated with increasing score. Table 3 demonstrates that for each doubling of the SCC the LS increases by one. For second lactation and older cows, each LS unit increase above LS 2.0 equals a loss of 200 kg of milk per lactation or 0.66 kg of milk per day. The milk yield loss of first lactation animals is estimated to be one-half that of older cows as shown in Fig. 1. The average LS gives a less distorted and more accurate picture of a lactation than does the average raw SCC.

Table 3 Comparative CMT score, SCC and linear scores (LS, also known as somatic cell score, SCS) and reduction in milk production as LS increases

Average linear score	CMT	SCC (cells/mL) *1000	Milk loss (%)	Milk loss per cow* per year		Milk loss per day*	
				Lb	Kg	Lb	Kg
2	Negative	50	Base	0	0	0	0
3	Negative	100	3	400	182	1.5	0.7
4	Negative	200	6	800	364	3	1.4
	Trace	300	7	1000	455		
5		400	8	1200	545	4.5	2.0
		500	9	1300	591		
	1	600	10	1400	636		
		700	10	1500	682		
6		800	11	1600	727	6	2.7
		900	11	1650	750		
		1000	12	1700	773		
	2	1200	>12	1700	773		
7		1600	>12	2000	909	7.5	3.4
8		3200	>12	2400	1091	9	4.1
9	3	6400	>12	2800	1273	10.5	4.8

* Loss in first lactation animals is 50% of the numbers indicated

Milk loss caused by pathogen-specific CM was studied extensively in two herds involving 3071 cows (Gröhn et al., 2004). The pathogens studied were *Streptococcus* spp., *Staphylococcus aureus*, *Staphylococcus* spp., *E. coli*, *Klebsiella* spp., *Arcanobacterium pyogenes* (now *Trueperella pyogenes*), other pathogens grouped together and 'no pathogen isolated'. Separate models were fitted for primiparous and multiparous because of the different shapes of their lactation curves. Lactation curves for major mastitis pathogens identified in multiparous cows can be seen in Fig. 2. It is important to note that cows with CM tended to be higher producers than their non-mastitic herdmates. Among older cows, *Streptococcus* spp., *Staph. aureus*, *T. pyogenes*, *E. coli* and *Klebsiella* spp. caused the most significant milk losses by multiparous cows. The results indicate that milk loss in mastitic cows did vary depending on the pathogen responsible for the mastitis. Among parity 1 cows, *Staph. aureus*, *E. coli* and *Klebsiella* spp. caused the greatest declines in milk yield. Days to the clinical event also varied by parity with more infections occurring earlier in lactation for parity 1 cows than multiparous cows. Milk yield also dropped in clinically mastitic cows for whom no pathogen was isolated. In another study (Hertl et al., 2011), *E. coli* was the most common pathogen in the initial CM case; however, the greatest milk production losses in multiparous animals were associated with *Klebsiella* spp. *Streptococcus* spp. occurred most frequently as the first CM case of primiparous cows, although *E. coli*

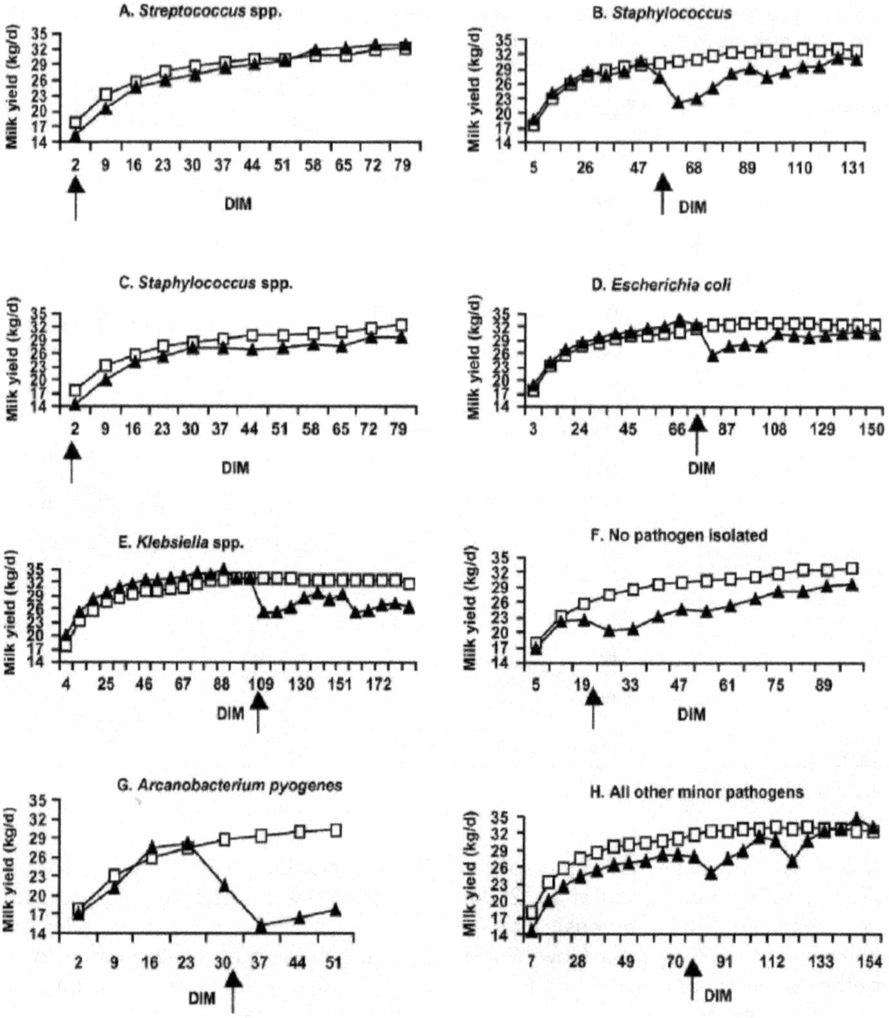

Figure 1 Impact of mastitis pathogens on milk loss for primiparous cows.

Y. T. Gröhn, D.J. Wilson, R.N. González, J.A. Hertl, H. Schulte, G. Bennett and Y.H. Schukken, 2004. Effect of Pathogen-Specific Clinical Mastitis on Milk Yield in Dairy Cows. *J. Dairy Sci.* 87: 3358–74. © American Dairy Science Association.

infection was associated with the greatest losses. Cha et al. (2011) demonstrated that the average cost per clinical case (US$) of Gram-positive, Gram-negative and other clinical mastitis were $133.73, $211.03 and $95.31, respectively. Treatment costs were the main contributor of the total costs associated with Gram-positive and 'other CM' causes (51.5% and 49.2%, respectively). Milk loss accounted for 72.4% of the cost associated with Gram-negative CM.

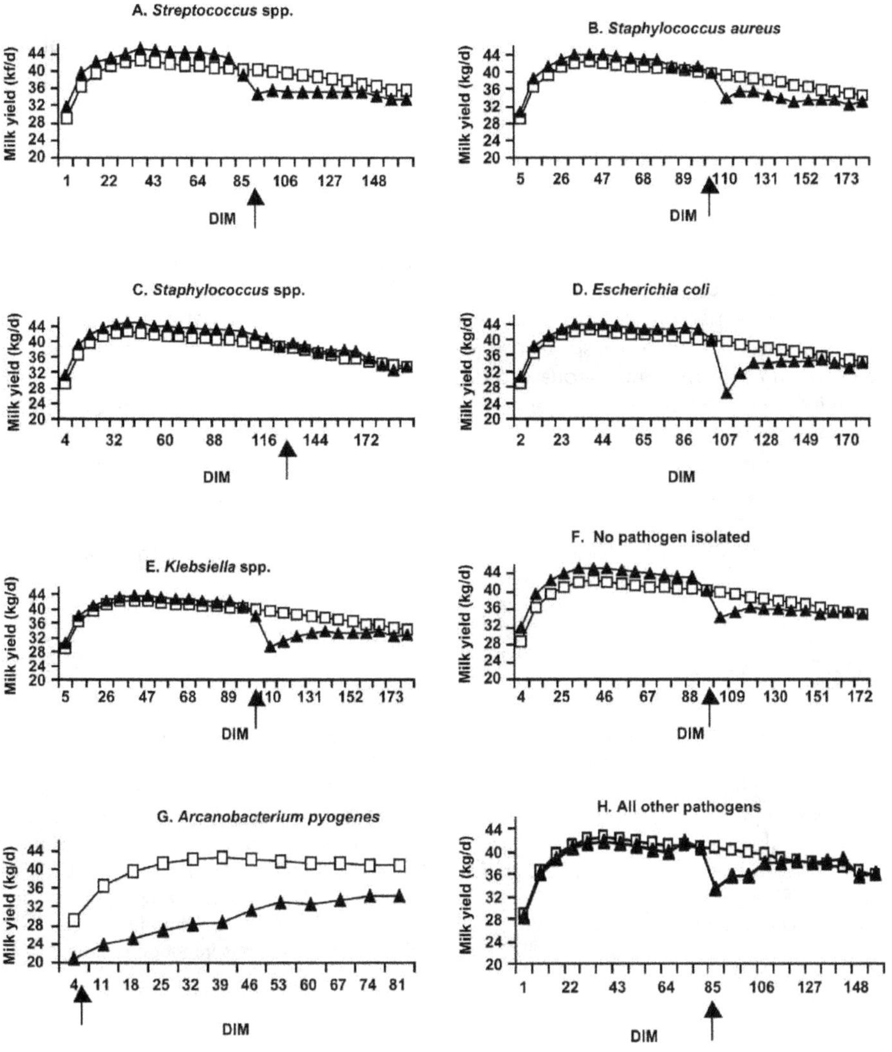

Figure 2 Impact of mastitis pathogens on milk loss for multiparous cows.

Y.T. Gröhn, D.J. Wilson, R.N. González, J.A. Hertl, H. Schulte, G. Bennett and Y.H. Schukken, 2004. Effect of Pathogen-Specific Clinical Mastitis on Milk Yield in Dairy Cows. *J. Dairy Sci.* 87: 3358–74. © American Dairy Science Association.

Many mastitis researchers have attributed clinical mastitis cases with 'no pathogen isolated' to Gram-negative pathogens that were eliminated by the cows' innate defence mechanisms. The similarities of the lactation curves and milk losses among 'no pathogen isolated', *E. coli* and *Klebsiella* spp. clinical infections appear to support this theory. In general, in both primiparous and multiparous cows, milk yield often began to drop several weeks before diagnosis of CM. The milk losses were greatest soon after diagnosis and then generally tapered off in subsequent weeks. The lower production level often continued for some time, and many CM-affected cows never regained their projected preclinical event milk yield.

6 Conclusion and future trends

Bovine mastitis is a common and complex disease involving many different management and environmental factors often unique to individual dairy farms. Environmental factors such as facilities design, management and sanitation impact infection risk as do milk harvesting systems, procedures and routines. Resident pathogens, contagious and environmental, and the farm's specific infection control efforts also influence infection risk. The characteristics of the numerous and distinct pathogens that infect the mammary gland greatly influence the productivity and profitability of the farm and the affected animals. Because mastitis treatments and control frequently involve intensive use of antimicrobials and the current public concern for reducing antimicrobial use by the dairy industry and veterinary profession, more efficient means of making treatment decisions while achieving and maintaining optimal cure rates must be developed. Using culture results and basing treatment protocols on pathogen identification and severity of inflammation are logical solutions to reduce antimicrobial use on dairy farms.

7 References

Abd El-Salam, M.H. 2014. Application of Proteomics to the areas of milk production, processing and quality control – A review. *Int. J. Dairy Technol.* 67, 12116

Addis, M.F., Pisanu, S., Ghisaura, S., Pagnozzi, D., Marogna, G., Tanca, A., Biosa, G., Cacciotto, C., Alberti, A., Pittau, M., Roggio, T. and Uzzau, S. 2011. Proteomics and pathway analyses of the milk fat globule in sheep naturally infected by Mycoplasma agalactiae provide indications of the in vivo response of the mammary epithelium to bacterial infection. *Infect. Immun.* 79, 3833–3845.

Addis, M.F., Pisanu, S., Marogna, G., Cubeddu, T., Pagnozzi, D., Cacciotto, C., Campesi, F., Schianchi, G., Rocca, S. and Uzzau, S. 2013. Production and release of antimicrobial and immune defense proteins by mammary epithelial cells following Streptococcus uberis infection of sheep. *Infect. Immun.* 81, 3182–3197.

Addis, M.F., Tedde, V., Dore, S., Pisanu, S., Puggioni, G.M.G., Roggio, A.M., Pagnozzi, D., Lollai, S., Cannas, E.A. and Uzzau, S., 2016. Evaluation of milk cathelicidin for detection of dairy sheep mastitis. *J. Dairy Sci.* 99, 6446–6456.

Addis, M.F., Tedde, V., Puggioni, G.M.G., Pisanu, S., Casula, A., Locatelli, C., Rota, N., Bronzo, V., Moroni, P. and Uzzau, S. 2016. Evaluation of milk cathelicidin for detection of bovine mastitis. *J. Dairy Sci.* 99, 8250–8258.

Åkerstedt, M., Waller, K.P., Larsen, L.B., Forsbäck, L. and Sternesjö, Å. 2008. Relationship between haptoglobin and serum amyloid A in milk and milk quality. *Int. Dairy J.* 18, 669–74.

Allen, J.C. 1990. Milk synthesis and secretion rates in cows with milk composition changed by oxytocin. *J. Dairy Sci.* 73, 975–984.

Anderson, K.L. and Walker, R.L. 1988. Sources of Prototheca spp in a dairy herd environment. *J. Am. Vet. Med. Assoc.* 193, 553–556.

Auldist, M.J., Coats, S., Rogers, G.L. and McDowell, G.H. 1995. Changes in the composition of milk from healthy and mastitic dairy cows during the lactation cycle. *Aust. J. Exp. Agric.* 35, 427–436.

Auldist, M.J., Coats, S., Sutherland, B.J., Mayes, J.J., McDowell, G.H. and Rogers, G.L. 1996. Effects of somatic cell count and stage of lactation on raw milk composition and the yield and quality of Cheddar cheese. *J. Dairy Res.* 63, 269–280.

Auldist, M.J. and Hubble, I.B. 1998. Effects of mastitis on raw milk and dairy products. *Aust. J. Dairy Technol.* 53, 28–36.

Babaei, H., Mansouri-Najand, L., Molaei, M.M., Kheradmand, A. and Sharifan, M. 2007. Assessment of lactate dehydrogenase, alkaline phosphatase and aspartate aminotransferase activities in cow's milk as an indicator of subclinical mastitis. *Vet. Res. Commun.* 31, 419–425.

Bar, D., Gröhn, Y.T., Bennett, G., González, R.N., Hertl, J.A., Schulte, H.F., Tauer, L.W., Welcome, F.L. and Schukken, Y.H. 2007. Effect of repeated episodes of generic clinical mastitis on milk yield in dairy cows. *J. Dairy Sci.* 90, 4643–4653.

Barbano, D.M., Ma, Y. and Santos, M. V. 2006. Influence of raw milk quality on fluid milk shelf life. *J. Dairy Sci.* 89 Suppl 1, E15–9.

Barkema, H.W., Schukken, Y.H., Lam, T.J., Galligan, D.T., Beiboer, M.L. and Brand, A. 1997. Estimation of interdependence among quarters of the bovine udder with subclinical mastitis and implications for analysis. *J. Dairy Sci.* 80, 1592–1599.

Bogin, E. and Ziv, G. 1973. Enzymes and minerals in normal and mastitic milk. *Cornell Vet.* 63, 666–676.

Bozzo, G., Bonerba, E., Di Pinto, A., Bolzoni, G., Ceci, E., Mottola, A., Tantillo, G. and Terio, V. 2014. Occurrence of Prototheca spp. in cow milk samples. *New Microbiol.* 37, 459–464.

Bradley, A. and Green, M. 2005. Use and interpretation of somatic cell count data in dairy cows. *In Pract.* 27, 310–315.

Bruckmaier, R.M., Ontsouka, C.E. and Blum, J.W. 2004. Fractionized milk composition in dairy cows with subclinical mastitis. *Vet. Med.* - UZPI (Czech Republic) 49, 283–290.

Cebra, C.K., Garry, F.B., Dinsmore, R.P. 1996. Naturally occurring acute coliform mastitis in Holstein cattle. *J. Vet. Intern. Med.* 10, 252–257.

Ceciliani, F., Ceron, J.J., Eckersall, P.D., Sauerwein, H. 2012. Acute phase proteins in ruminants. *J. Proteomics* 75, 4207–4231.

Ceciliani, F., Eckersall, D., Burchmore, R., Lecchi, C., 2014. Proteomics in veterinary medicine: applications and trends in disease pathogenesis and diagnostics. *Vet. Pathol.* 51, 351–362.

Cha, E., Bar, D., Hertl, J.A., Tauer, L.W., Bennett, G., González, R.N., Schukken, Y.H., Welcome, F.L., Gröhn, Y.T. 2011. The cost and management of different types of clinical mastitis in dairy cows estimated by dynamic programming. *J. Dairy Sci.* 94, 4476–4487.

Cha, E., Hertl, J.A., Schukken, Y.H., Tauer, L.W., Welcome, F.L., Gröhn, Y.T. 2013. The effect of repeated episodes of bacteria-specific clinical mastitis on mortality and culling in Holstein dairy cows. *J. Dairy Sci.* 96, 4993–5007.

Chagunda, M.G., Larsen, T., Bjerring, M., Ingvartsen, K.L. 2006. L-lactate dehydrogenase and N-acetyl-beta-D-glucosaminidase activities in bovine milk as indicators of non-specific mastitis. *J. Dairy Res.* 73, 431–440.

Chahota, R., Katoch, R., Mahajan, A., Verma, S. 2001. Clinical bovine mastitis caused by Geotrichum candidum. *Vet. Ark.* 71, 197–201.

Chen, M.-H., Hung, S.-W., Shyu, C.-L., Lin, C.-C., Liu, P.-C., Chang, C.-H., Shia, W.-Y., Cheng, C.-F., Lin, S.-L., Tu, C.-Y., Lin, Y.-H., Wang, W.-S. 2012. Lactococcus lactis subsp. lactis infection in Bester sturgeon, a cultured hybrid of Huso huso × Acipenser ruthenus, in Taiwan. *Res. Vet. Sci.* 93, 581–588.

Cook, N.B., Bennett, T.B., Emery, K.M., Nordlund, K. V. 2002. Monitoring nonlactating cow intramammary infection dynamics using DHI somatic cell count data. *J. Dairy Sci.* 85, 1119–1126.

Corbellini, L.G., Driemeier, D., Cruz, C., Dias, M.M. and Ferreiro, L. 2001. Bovine mastitis due to Prototheca zopfii: clinical, epidemiological and pathological aspects in a Brazilian dairy herd. *Trop. Anim. Health Prod.* 33, 463–470.

Costa, E.O., Melville, P.A., Ribeiro, A.R., Watanabe, E.T. and Parolari, M.C. 1997. Epidemiologic study of environmental sources in a Prototheca zopfii outbreak of bovine mastitis. *Mycopathologia* 137, 33–36.

Davis, S.R., Farr, V.C., Prosser, C.G., Nicholas, G.D., Turner, S.-A., Lee, J., Hart, A.L. 2004. Milk L-lactate concentration is increased during mastitis. *J. Dairy Res.* 71, 175–181.

De Vliegher, S., Fox, L.K., Piepers, S., McDougall, S., Barkema, H.W. 2012. Invited review: Mastitis in dairy heifers: nature of the disease, potential impact, prevention, and control. *J. Dairy Sci.* 95, 1025–1040.

Deluyker, H.A., Van Oye, S.N., Boucher, J.F. 2005. Factors affecting cure and somatic cell count after pirlimycin treatment of subclinical mastitis in lactating cows. *J. Dairy Sci.* 88, 604–614.

Devriese, L.A., Hommez, J., Laevens, H., Pot, B., Vandamme, P., Haesebrouck, F. 1999. Identification of aesculin-hydrolyzing streptococci, lactococci, aerococci and enterococci from subclinical intramammary infections in dairy cows. *Vet. Microbiol.* 70, 87–94.

Devriese, L.A., Schleifer, K.H., Adegoke, G.O. 1985. Identification of coagulase-negative staphylococci from farm animals. *J. Appl. Bacteriol.* 58, 45–55.

Djabri, B., Bareille, N., Beaudeau, F., Seegers, H. 2002. Quarter milk somatic cell count in infected dairy cows: A meta-analysis. *Vet. Res.* 33, 335–357

Dohoo, I.R., Leslie, K.E. 1991. Evaluation of changes in somatic cell counts as indicators of new intramammary infections. *Prev. Vet. Med.* 10, 225–237.

Dohoo, I.R., Meek, A.H. 1982. Somatic cell counts in bovine milk. *Can. Vet. journal. La Rev. vétérinaire Can.* 23, 119–125.

Eaton, J.W., Brandt, P., Mahoney, J.R., Lee, J.T. 1982. Haptoglobin: a natural bacteriostat. *Science* 215, 691–693.

Eckersall, P.D., Young, F.J., McComb, C., Hogarth, C.J., Safi, S., Weber, A., McDonald, T., Nolan, A.M., Fitzpatrick, J.L. 2001. Acute phase proteins in serum and milk from dairy cows with clinical mastitis. *Vet. Rec.* 148, 35–41.

Eriksson, Å., Persson Waller, K., Svennersten-Sjaunja, K., Haugen, J.-E., Lundby, F., Lind, O., 2005. Detection of mastitic milk using a gas-sensor array system (electronic nose). *Int. Dairy J.* 15, 1193–1201.

Facklam, R., Elliott, J.A. 1995. Identification, classification, and clinical relevance of catalase-negative, gram-positive cocci, excluding the streptococci and enterococci. *Clin. Microbiol. Rev.* 8, 479–495.

Fernandes, A.M., Oliveira, C.A.F., Lima, C.G. 2007. Effects of somatic cell counts in milk on physical and chemical characteristics of yoghurt. *Int. Dairy J.* 17, 111–115.

Forsbäck, L., Lindmark-Månsson, H., Andrén, A., Svennersten-Sjaunja, K. 2010. Evaluation of quality changes in udder quarter milk from cows with low-to-moderate somatic cell counts. *Animal* 4, 617–626.

Fortin, M., Messier, S., Paré, J., Higgins, R. 2003. Identification of catalase-negative, non-beta-hemolytic, gram-positive cocci isolated from milk samples. *J. Clin. Microbiol.* 41, 106–109.

Fry, P.R., Middleton, J.R., Dufour, S., Perry, J., Scholl, D., Dohoo, I. 2014. Association of coagulase-negative staphylococcal species, mammary quarter milk somatic cell count, and persistence of intramammary infection in dairy cattle. *J. Dairy Sci.* 97, 4876–4885.

Gao, J., Zhang, H., He, J., He, Y., Li, S., Hou, R., Wu, Q., Gao, Y., Han, B., 2012. Characterization of Prototheca zopfii associated with outbreak of bovine clinical mastitis in herd of Beijing, China. *Mycopathologia* 173, 275–281.

Gordoncillo, M.J.N., Bautista, J.A.N., Hikiba, M., Sarmago, I.G., Haguingan, J.M.B. 2010. Comparison of conventionally identified mastitis bacterial organisms with commercially available microbial identification kit (BBL Crystal ID®). *Philipp. J. Vet. Med.* 47, 54–57.

Gröhn, Y.T., Wilson, D.J., González, R.N., Hertl, J.A., Schulte, H., Bennett, G., Schukken, Y.H. 2004. Effect of pathogen-specific clinical mastitis on milk yield in dairy cows. *J. Dairy Sci.* 87, 3358–3374.

Guélat-Brechbuehl, M., Thomann, A., Albini, S., Moret-Stalder, S., Reist, M., Bodmer, M., Michel, A., Niederberger, M.D., Kaufmann, T. 2010. Cross-sectional study of Streptococcus species in quarter milk samples of dairy cows in the canton of Bern, Switzerland. *Vet. Rec.* 167, 211–215.

Gurjar, A., Gioia, G., Schukken, Y., Welcome, F., Zadoks, R., Moroni, P. 2012. Molecular diagnostics applied to mastitis problems on dairy farms. *Vet. Clin. North Am. Food Anim. Pract.* 28, 565–576.

Haddadi, K., Moussaoui, F., Hebia, I., Laurent, F., Le Roux, Y. 2005. E. coli proteolytic activity in milk and casein breakdown. *Reprod. Nutr. Dev.* 45, 485–496.

Hamann, J. 2005. Diagnosis of mastitis and indicators of milk quality, Mastitis in Dairy Production; Current Knowledge and Future Solutions. Wageningen Academic Publishers, Wageningen, The Netherlands.

Hamann, J. 1996. Somatic cells: factors of influence and practical measures to keep a physiological level. *Mastit. Newsl.* 21, 9–11.

Harmon, R.J., 1994. Physiology of mastitis and factors affecting somatic cell counts. *J. Dairy Sci.* 77, 2103–12. doi:10.3168/jds.S0022-0302(94)77153-8

Harrison, E., Bonhotal, J., M., S. 2008. Using manure solids as bedding. Final Report. Cornell University. Waste Management Institute, Ithaca, NY.

Hertl, J. a, Gröhn, Y.T., Leach, J.D.G., Bar, D., Bennett, G.J., González, R.N., Rauch, B.J., Welcome, F.L., Tauer, L.W., Schukken, Y.H. 2010. Effects of clinical mastitis caused by gram-positive and gram-negative bacteria and other organisms on the probability of conception in New York State Holstein dairy cows. *J. Dairy Sci.* 93, 1551–1560.

Hertl, J.A., Schukken, Y.H., Bar, D., Bennett, G.J., González, R.N., Rauch, B.J., Welcome, F.L., Tauer, L.W., Gröhn, Y.T. 2011. The effect of recurrent episodes of clinical mastitis caused by gram-positive and gram-negative bacteria and other organisms on mortality and culling in Holstein dairy cows. *J. Dairy Sci.* 94, 4863–4877.

Hertl, J.A., Schukken, Y.H., Welcome, F.L., Tauer, L.W., Gröhn, Y.T. 2014. Pathogen-specific effects on milk yield in repeated clinical mastitis episodes in Holstein dairy cows. *J. Dairy Sci.* 97: 1465–1480.

Hiss, S., Mielenz, M., Bruckmaier, R.M., Sauerwein, H. 2004. Haptoglobin concentrations in blood and milk after endotoxin challenge and quantification of mammary Hp mRNA expression. *J. Dairy Sci.* 87, 3778–3784.

Hiss, S., Mueller, U., Neu-Zahren, A., Sauerwein, H. 2007. Haptoglobin and lactate dehydrogenase measurements in milk for the identification of subclinically diseased udder quarters. *Vet. Med. (Praha).* 52, 245–252.

Hogan, J.S., Smith, K.L. 1997. Occurrence of clinical and subclinical environmental streptococcal mastitis, in *Proceedings of the Symposium on Udder Health Management for Environmental Streptococci.* Ontario Veterinary College, Canada, pp. 36–41.

Holdaway, R.J. 1990. A comparison of methods for the diagnosis of bovine subclinical mastitis within New Zealand dairy herds. PhD Thesis.

Holdaway, R.J., Holmes, C.W., Steffert, I.J. 1996. A comparison of indirect methods for diagnosis of subclinical intramammary infection in lactating dairy cows. Part 2: the discriminative ability of eight parameters in foremilk from individual quarters and cows. *Aust. J. Dairy Tech.* 51, 72–78.

Holm, C., Jepsen, L., Larsen, M., Jespersen, L. 2004. Predominant microflora of downgraded Danish bulk tank milk. *J. Dairy Sci.* 87, 1151–1157.

Hovinen, M., Aisla, A.-M., Pyörälä, S. 2006. Accuracy and reliability of mastitis detection with electrical conductivity and milk colour measurement in automatic milking. *Acta Agric. Scand. Sect. A - Anim. Sci.* 56, 121–127.

Huerre, M., Ravisse, P., Solomon, H., Ave, P., Briquelet, N., Maurin, S., Wuscher, N. 1993. [Human protothecosis and environment]. *Bull. la Société Pathol. Exot.* 86, 484–488.

Husfeldt, A.W., Endres, M.I., Salfer, J.A., Janni, K.A. 2012. Management and characteristics of recycled manure solids used for bedding in Midwest freestall dairy herds. *J. Dairy Sci.* 95, 2195–2203.

Ibeagha-awemu, E.M., Ibeagha, A.E., Messier, S., Zhao, X. 2010. Proteomics, genomics, and pathway analyses of Escherichia coli and Staphylococcus aureus infected milk whey reveal molecular pathways and networks involved in mastitis. *J. Proteome Res.* 4604–4619.

Jagielski, T., Lassa, H., Ahrholdt, J., Malinowski, E., Roesler, U. 2011. Genotyping of bovine Prototheca mastitis isolates from Poland. *Vet. Microbiol.* 149, 283–287.

Jánosi, S., Rátz, F., Szigeti, G., Kulcsár, M., Kerényi, J., Laukó, T., Katona, F., Huszenicza, G. 2001. Review of the microbiological, pathological, and clinical aspects of bovine mastitis caused by the alga Prototheca zopfii. *Vet. Q.* 23, 58–61.

Jayarao, B.M., Oliver, S.P., Tagg, J.R., Matthews, K.R. 1991. Genotypic and phenotypic analysis of Streptococcus uberis isolated from bovine mammary secretions. *Epidemiol. Infect.* 107, 543–555.

Karlsson, A., Arvidson, S. 2002. Variation in extracellular protease production among clinical isolates of Staphylococcus aureus due to different levels of expression of the protease repressor sarA. *Infect. Immun.* 70, 4239–4246.

Katholm, J., Andersen, P.H. 1992. Acute coliform mastitis in dairy cows: endotoxin and biochemical changes in plasma and colony-forming units in milk. *Vet. Rec.* 131, 513–514.

Kishimoto, Y., Kano, R., Maruyama, H., Onozaki, M., Makimura, K., Ito, T., Matsubara, K., Hasegawa, A., Kamata, H. 2010. 26S rDNA-based phylogenetic investigation of Japanese cattle-associated Prototheca zopfii isolates. *J. Vet. Med. Sci.* 72, 123–126.

Kitchen, B.J. 1981. Review of the progress of dairy science: bovine mastitis: milk compositional changes and related diagnostic tests. *J. Dairy Res.* 48, 167–188.

Kossaibati, M.A., Esslemont, R.J. 1997. The costs of production diseases in dairy herds in England. *Vet. J.* 154, 41–51.

Kuang, Y., Tani, K., Synnott, A.J., Ohshima, K., Higuchi, H., Nagahata, H., Tanji, Y. 2009. Characterization of bacterial population of raw milk from bovine mastitis by culture-independent PCR–DGGE method. *Biochem. Eng. J.* 45, 76–81.

Laevens, H., Deluyker, H., Schukken, Y.H., De Meulemeester, L., Vandermeersch, R., De Muêlenaere, E., De Kruif, A. 1997. Influence of parity and stage of lactation on the somatic cell count in bacteriologically negative dairy cows. *J. Dairy Sci.* 80, 3219–3226.

Larson, M.A., Weber, A., Weber, A.T., McDonald, T.L. 2005. Differential expression and secretion of bovine serum amyloid A3 (SAA3) by mammary epithelial cells stimulated with prolactin or lipopolysaccharide. *Vet. Immunol. Immunopathol.* 107, 255–264.

Lass-Flörl, C., Mayr, A. 2007. Human protothecosis. *Clin. Microbiol. Rev.* 20, 230–242.

Le Maréchal, C., Thiéry, R., Vautor, E., Le Loir, Y. 2011. Mastitis impact on technological properties of milk and quality of milk products—a review. *Dairy Sci. Technol.* 91, 247–282

Lopez-Benavides, M.G., Dohoo, I., Scholl, D., Middleton, J.R., Perez, R. 2012. Interpreting Bacteriological Culture Results to Diagnose Bovine Intramammary Infections. *Natl. Mastit. Counc. Res. Comm. Rep.*

Lopez-Benavides, M.G., Williamson, J.H., Pullinger, G.D., Lacy-Hulbert, S.J., Cursons, R.T., Leigh, J.A. 2007. Field observations on the variation of Streptococcus uberis populations in a pasture-based dairy farm. *J. Dairy Sci.* 90, 5558–5566.

Malek dos Reis, C.B., Barreiro, J.R., Mestieri, L., Porcionato, M.A. de F., dos Santos, M.V. 2013. Effect of somatic cell count and mastitis pathogens on milk composition in Gyr cows. *BMC Vet. Res.* 9, 67.

Malinowski, E., Kłossowska, A., Kaczmarowski, M., Kuźma, K., Markiewicz, H. 2003. Field trials on the prophylaxis of intramammary infections in pregnant heifers. *Pol. J. Vet. Sci.* 6, 117–124.

Marques, S., Silva, E., Kraft, C., Carvalheira, J., Videira, A., Huss, V.A.R., Thompson, G. 2008. Bovine mastitis associated with Prototheca blaschkeae. *J. Clin. Microbiol.* 46, 1941–1945.

Mattila, T., Pyörälä, S., Sandholm, M. 1986. Comparison of milk antitrypsin, albumin, n-acetyl-beta-D-glucosaminidase, somatic cells and bacteriological analysis as indicators of bovine subclinical mastitis. *Vet. Res. Commun.* 10, 113–124.

McDonald, T.L., Larson, M.A., Mack, D.R., Weber, A. 2001. Elevated extrahepatic expression and secretion of mammary-associated serum amyloid A 3 (M-SAA3) into colostrum. *Vet. Immunol. Immunopathol.* 83, 203–211.

Merin, U., Fleminger, G., Komanovsky, J., Silanikove, N., Bernstein, S., Leitner, G. 2008. Subclinical udder infection with *Streptococcus dysgalactiae* impairs milk coagulation properties: The emerging role of proteose peptones. *Dairy Sci. Technol.* 88, 407–419.

Miglio, A., Moscati, L., Fruganti, G., Pela, M., Scoccia, E., Valiani, A., Maresca, C. 2013. Use of milk amyloid A in the diagnosis of subclinical mastitis in dairy ewes. *J. Dairy Res.* 80, 496–502.

Mitchell, G.E., Fedrick, I.A., Rogers, S.A. 1986. The relationship between somatic cell count, composition and manufacturing properties of bulk milk. I. Composition of farm bulk milk. *Aust. J. Dairy Technol.* 41.9–12.

Möller, A., Truyen, U., Roesler, U. 2007. Prototheca zopfii genotype 2: the causative agent of bovine protothecal mastitis? *Vet. Microbiol.* 120, 370–374.

Mottram, T., Rudnitskaya, A., Legin, A., Fitzpatrick, J.L., Eckersall, P.D. 2007. Evaluation of a novel chemical sensor system to detect clinical mastitis in bovine milk. *Biosens. Bioelectron.* 22, 2689–2693.

Murakami, M., Dorschner, R.A., Stern, L.J., Lin, K.H., Gallo, R.L. 2005. Expression and Secretion of Cathelicidin Antimicrobial Peptides in Murine Mammary Glands and Human Milk. *Pediatr. Res.* 57, 10–15.

Nguyen, D.A., Neville, M.C. 1998. Tight junction regulation in the mammary gland. *J. Mammary Gland Biol. Neoplasia* 3, 233–246.

Nielsen, N.I., Larsen, T., Bjerring, M., Ingvartsen, K.L. 2005. Quarter health, milking interval, and sampling time during milking affect the concentration of milk constituents. *J. Dairy Sci.* 88, 3186–3200.

Nomura, M., Kobayashi, M., Narita, T., Kimoto-Nira, H., Okamoto, T. 2006. Phenotypic and molecular characterization of *Lactococcus lactis* from milk and plants. *J. Appl. Microbiol.* 101, 396–405.

Norberg, E. 2005. Electrical conductivity of milk as a phenotypic and genetic indicator of bovine mastitis: A review. *Livest. Prod. Sci.* 96, 129–139.

O'Mahony, M.C., Healy, A.M., Harte, D., Walshe, K.G., Torgerson, P.R., Doherty, M.L. 2006. Milk amyloid A: correlation with cellular indices of mammary inflammation in cows with normal and raised serum amyloid A. *Res. Vet. Sci.* 80, 155–161.

Odierno, L., Calvinho, L., Traverssa, P., Lasagno, M., Bogni, C., Reinoso, E., 2006. Conventional identification of *Streptococcus uberis* isolated from bovine mastitis in Argentinean dairy herds. *J. Dairy Sci.* 89, 3886–3890.

Oliveira, C.A.F., Fernandes, A.M., Neto, O.C.C., Fonseca, L.F.L., Silva, E.O.T., Balian, S.C. 2002. Composition and sensory evaluation of whole yogurt produced from milk with different somatic cell counts. *Aust. J. Dairy Technol.* 57, 192–196.

Oliver, S.P., Murinda, S.E. 2012. Antimicrobial resistance of mastitis pathogens. *Vet. Clin. North Am. Food Anim. Pract.* 28, 165–185.

Pemberton, R.M., Hart, J.P., Mottram, T.T. 2001. An assay for the enzyme N-acetyl-beta-D-glucosaminidase (NAGase) based on electrochemical detection using screen-printed carbon electrodes (SPCEs). *Analyst* 126, 1866–1871.

Pieper, L., Godkin, A., Roesler, U., Polleichtner, A., Slavic, D., Leslie, K.E., Kelton, D.F. 2012. Herd characteristics and cow-level factors associated with Prototheca mastitis on dairy farms in Ontario, Canada. *J. Dairy Sci.* 95, 5635–5644.

Piepers, S., Schukken, Y.H., Passchyn, P., De Vliegher, S. 2013. The effect of intramammary infection with coagulase-negative staphylococci in early lactating heifers on milk yield throughout first lactation revisited. *J. Dairy Sci.* 96, 5095–5105.

Piessens, V., Van Coillie, E., Verbist, B., Supré, K., Braem, G., Van Nuffel, A., De Vuyst, L., Heyndrickx, M., De Vliegher, S. 2011. Distribution of coagulase-negative Staphylococcus species from milk and environment of dairy cows differs between herds. *J. Dairy Sci.* 94, 2933–2944.

Plumed-Ferrer, C., Gazzola, S., Fontana, C., Bassi, D., Cocconcelli, P., Wright, V. 2015. Genome sequence of *Lactococcus lactis* subsp. *cremoris* Mast36, a strain Isolated from Bovine Mastitis. *Genome Announc.* 3, 5–6.

Plumed-Ferrer, C., Uusikylä, K., Korhonen, J., von Wright, A. 2013. Characterization of *Lactococcus lactis* isolates from bovine mastitis. *Vet. Microbiol.* 167, 592–599.

Pot, B., Devriese, L.A., Ursi, D., Vandamme, P., Haesebrouck, F., Kersters, K. 1996. Phenotypic identification and differentiation of Lactococcus strains isolated from animals. *Syst. Appl. Microbiol.* 19, 213–222.

Pryor, S.M., Cursons, R.T., Williamson, J.H., Lacy-Hulbert, S.J. 2009. Experimentally induced intramammary infection with multiple strains of *Streptococcus uberis*. *J. Dairy Sci.* 92, 5467–5475.

Pyorala, S. 2003. Indicators of inflammation in the diagnosis of mastitis. *Vet. Res.* 34, 565–578.
Reinhardt, T.A., Sacco, R.E., Nonnecke, B.J., Lippolis, J.D. 2013. Bovine milk proteome: quantitative changes in normal milk exosomes, milk fat globule membranes and whey proteomes resulting from Staphylococcus aureus mastitis. *J. Proteomics* 82, 141–154.
Rhoda, D.A., Pantoja, J.C.F. 2012. Using mastitis records and somatic cell count data. *Vet. Clin. North Am. Food Anim. Pract.* 28, 347–361.
Ricchi, M., De Cicco, C., Buzzini, P., Cammi, G., Arrigoni, N., Cammi, M., Garbarino, C. 2013. First outbreak of bovine mastitis caused by Prototheca blaschkeae. *Vet. Microbiol.* 162, 997–999.
Ricchi, M., Goretti, M., Branda, E., Cammi, G., Garbarino, C.A., Turchetti, B., Moroni, P., Arrigoni, N., Buzzini, P. 2010. Molecular characterization of Prototheca strains isolated from Italian dairy herds. *J. Dairy Sci.* 93, 4625–4631.
Roberson, J.R., Warnick, L.D., Moore, G. 2004. Mild to moderate clinical mastitis: efficacy of intramammary amoxicillin, frequent milk-out, a combined intramammary amoxicillin, and frequent milk-out treatment versus no treatment. *J. Dairy Sci.* 87, 583–592
Roesler, U., Hensel, A. 2003. Longitudinal analysis of Prototheca zopfii-specific immune responses: correlation with disease progression and carriage in dairy cows. *J. Clin. Microbiol.* 41, 1181–1186.
Roesler, U., Scholz, H., Hensel, A. 2001. Immunodiagnostic identification of dairy cows infected with Prototheca zopfii at various clinical stages and discrimination between infected and uninfected cows. *J. Clin. Microbiol.* 39, 539–543.
Rogers, S.A., Mitchell, G.E., Bartley, J.P. 1994. The relationship between somatic cell count, composition and manufacturing properties of bulk milk 6. Cheddar cheese and skim milk yoghurt. *Aust. J. Dairy Technol.* 49, 70–74.
Ruegg, P.L. 2003. Investigation of mastitis problems on farms. *Vet. Clin. North Am. Food Anim. Pract.* 19, 47–73.
Ruegg, P.L., Pantoja, J.C.F. 2013. Understanding and using somatic cell counts to improve milk quality. *Irish J. Agric. Food Res.* 52, 101–117.
Santos, J.E.P., Juchem, S.O., Cerri, R.L.A., Galvão, K.N., Chebel, R.C., Thatcher, W.W., Dei, C.S., Bilby, C.R. 2004. Effect of bST and reproductive management on reproductive performance of Holstein dairy cows. *J. Dairy Sci.* 87, 868–881.
Scaccabarozzi, L., Locatelli, C., Pisoni, G., Manarolla, G., Casula, A., Bronzo, V., Moroni, P. 2011. Short communication: Epidemiology and genotyping of Candida rugosa strains responsible for persistent intramammary infections in dairy cows. *J. Dairy Sci.* 94, 4574–4577.
Schepers, A.J., Lam, T.J., Schukken, Y.H., Wilmink, J.B., Hanekamp, W.J. 1997. Estimation of variance components for somatic cell counts to determine thresholds for uninfected quarters. *J. Dairy Sci.* 80, 1833–1840.
Schukken, Y.H., Bennett, G.J., Zurakowski, M.J., Sharkey, H.L., Rauch, B.J., Thomas, M.J., Ceglowski, B., Saltman, R.L., Belomestnykh, N., Zadoks, R.N. 2011. Randomized clinical trial to evaluate the efficacy of a 5-day ceftiofur hydrochloride intramammary treatment on nonsevere gram-negative clinical mastitis. *J. Dairy Sci.* 94, 6203–6215.
Schukken, Y.H., Hertl, J., Bar, D., Bennett, G.J., González, R.N., Rauch, B.J., Santisteban, C., Schulte, H.F., Tauer, L., Welcome, F.L., Gröhn, Y.T. 2009. Effects of repeated gram-positive and gram-negative clinical mastitis episodes on milk yield loss in Holstein dairy cows. *J. Dairy Sci.* 92, 3091–3105.
Schukken, Y.H., Wilson, D.J., Welcome, F., Garrison-Tikofsky, L., González, R.N. 2003. Monitoring udder health and milk quality using somatic cell counts. *Vet. Res.* 34, 579–596.
Sharma, K.K., Randolph, H.E. 1974. Influence of mastitis on properties of milk. 8. Distribution of soluble and micellar casein. *J. Dairy Sci.* 57, 19–23.
Shuster, D.E., Harmon, R.J., Jackson, J.A., Hemken, R.W. 1991. Suppression of milk production during endotoxin-induced mastitis. *J. Dairy Sci.* 74, 3763–3774.
Smolenski, G. a, Wieliczko, R.J., Pryor, S.M., Broadhurst, M.K., Wheeler, T.T., Haigh, B.J. 2011. The abundance of milk cathelicidin proteins during bovine mastitis. *Vet. Immunol. Immunopathol.* 143, 125–130.

Smolenski, G., Haines, S., Kwan, F.Y.S., Bond, J., Farr, V., Davis, S.R., Stelwagen, K., Wheeler, T.T. 2007. Characterisation of host defense proteins in milk using a proteomic approach. *J. Proteome Res.* 6, 207–215.

Smolenski, G.A., Broadhurst, M.K., Stelwagen, K., Haigh, B.J., Wheeler, T.T. 2014. Host defence related responses in bovine milk during an experimentally induced *Streptococcus uberis* infection. *Proteome Sci.* 12, 19.

Spanamberg, A., Wünder, E.A., Brayer Pereira, D.I., Argenta, J., Cavallini Sanches, E.M., Valente, P., Ferreiro, L. 2008. Diversity of yeasts from bovine mastitis in Southern Brazil. *Rev. Iberoam. Micol.* 25, 154–156.

Supré, K., Haesebrouck, F., Zadoks, R.N., Vaneechoutte, M., Piepers, S., De Vliegher, S. 2011. Some coagulase-negative Staphylococcus species affect udder health more than others. *J. Dairy Sci.* 94, 2329–2340.

Svec, P., Sedlácek, I. 2008. Characterization of Lactococcus lactis subsp. lactis isolated from surface waters. *Folia Microbiol. (Praha).* 53, 53–56.

Teixeira, L.M., Merquior, V.L., Vianni, M.C., Carvalho, M.G., Fracalanzza, S.E., Steigerwalt, A.G., Brenner, D.J., Facklam, R.R. 1996. Phenotypic and genotypic characterization of atypical Lactococcus garvieae strains isolated from water buffalos with subclinical mastitis and confirmation of L. garvieae as a senior subjective synonym of Enterococcus seriolicida. *Int. J. Syst. Bacteriol.* 46, 664–668.

Tenhagen, B., Kalbe, P., Baumgartner, B., Heuwieser, W. 2001. An outbreak of mastitis caused by Prototheca zopfii on a large confinement dairy: Analysis of cow level risk factors. *Proc. 2nd Int. Symp. Mastit. Milk Qual*, Vancouver, BC, Canada, 208–12.

Thompson, G., Silva, E., Marques, S., Müller, A., Carvalheira, J. 2009. Algaemia in a dairy cow by Prototheca blaschkeae. *Med. Mycol.* 47, 527–531.

Todhunter, D.A., Smith, K.L., Hogan, J.S. 1995. Environmental streptococcal intramammary infections of the bovine mammary gland. *J. Dairy Sci.* 78, 2366–2374.

Vanderhaeghen, W., Piepers, S., Leroy, F., Van Coillie, E., Haesebrouck, F., De Vliegher, S. 2015. Identification, typing, ecology and epidemiology of coagulase negative staphylococci associated with ruminants. *Vet. J.* 203, 44–51. doi:10.1016/j.tvjl.2014.11.001

Vanderhaeghen, W., Piepers, S., Leroy, F., Van Coillie, E., Haesebrouck, F., De Vliegher, S. 2014. Invited review: effect, persistence, and virulence of coagulase-negative Staphylococcus species associated with ruminant udder health. *J. Dairy Sci.* 97, 5275–5293.

Viguier, C., Arora, S., Gilmartin, N., Welbeck, K., O'Kennedy, R. 2009. Mastitis detection: current trends and future perspectives. *Trends Biotechnol.* 27, 486–493

Wenz, J.R., Barrington, G.M., Garry, F.B., McSweeney, K.D., Dinsmore, R.P., Goodell, G., Callan, R.J. 2001. Bacteremia associated with naturally occurring acute coliform mastitis in dairy cows. *J. Am. Vet. Med. Assoc.* 219, 976–981.

Wenz, J.R., Garry, F.B., Lombard, J.E., Elia, R., Prentice, D., Dinsmore, R.P. 2005. Short communication: Efficacy of parenteral ceftiofur for treatment of systemically mild clinical mastitis in dairy cattle. *J. Dairy Sci.* 88, 3496–3499.

Werner, B., Moroni, P., Gioia, G., Lavín-Alconero, L., Yousaf, A., Charter, M.E., Carter, B.M., Bennett, J., Nydam, D. V, Welcome, F., Schukken, Y.H. 2014. Short communication: Genotypic and phenotypic identification of environmental streptococci and association of *Lactococcus lactis* ssp. *lactis* with intramammary infections among different dairy farms. *J. Dairy Sci.* 97, 6964–6969.

Wheeler, T.T., Smolenski, G.A., Harris, D.P., Gupta, S.K., Haigh, B.J., Broadhurst, M.K., Molenaar, A.J., Stelwagen, K. 2012. Host-defence-related proteins in cows' milk. *Animal* 6, 415–422.

Whyte, D., Walmsley, M., Liew, A., Claycomb, R., Mein, G. 2005. Chemical and rheological aspects of gel formation in the California Mastitis Test. *J. Dairy Res.* 72, 115–121.

Wiesner, J., Vilcinskas, A. 2010. Antimicrobial peptides – The ancient arm of the human immune system. *Virulence* 1, 440–464.

Zadoks, R.N., Fitzpatrick, J.I. 2009. Changing trends in mastitis. *Ir. Vet. J.* 62 (Suppl 4), 59–70.

Zanetti, M. 2005. The role of cathelicidins in the innate host defenses of mammals. *Curr. Issues Mol. Biol.* 7, 179–196.

Zanetti, M. 2004. Cathelicidins, multifunctional peptides of the innate immunity. *J. Leukoc. Biol.* 75, 39–48.

Zaragoza, C.S., Olivares, R.A.C., Watty, A.E.D., Moctezuma, A. de la P., Tanaca, L.V. 2011. Yeasts isolation from bovine mammary glands under different mastitis status in the Mexican High Plateu. *Rev. Iberoam. Micol.* 28, 79–82.

Zhang, L., Boeren, S., van Hooijdonk, a. C.M., Vervoort, J.M., Hettinga, K. A. 2015. A proteomic perspective on the changes in milk proteins due to high somatic cell count. *J. Dairy Sci.* 98, 5339–5351.

Zhang, S., Maddox, C.W. 2000. Cytotoxic activity of coagulase-negative staphylococci in bovine mastitis. *Infect. Immun.* 68, 1102–1108.

Chapter 3

Advances in dairy cattle breeding to improve resistance to mastitis

John Cole, USDA-ARS, USA

1 Introduction

Mastitis, an inflammation of the mammary gland associated with bacterial infections, is generally regarded as the most costly disease of dairy cattle because of its high incidence and effects on milk production and composition (Seegers et al., 2003). Genetic selection for highly productive dairy cows has been very successful; however, udder health has declined in many dairy breeds because of its unfavourable correlations with milk production. Poor udder health results in higher veterinary and farm labour costs, increased rates of involuntary culling, decreased farm revenue and adverse impacts on animal welfare. However, genetic selection can be used to improve udder health just as it has been used to increase production (e.g. Schutz, 1994; Heringstad et al., 2003). Selection may be based on direct (e.g. cases of clinical infection) or indirect (e.g. somatic cell counts) indicators of mastitis. Genetic improvement programmes for resistance to clinical mastitis have often been limited to selection for improved somatic cell count (SCC) (or functions thereof), rather than records of clinical cases of disease, due to the cost of data collection. The now-routine use of on-farm computer for record-keeping and data transmission has increased the ease of data collection in many countries that previously did not record those data in a central database (e.g. Zottl, 2016).

Several new phenotypes that can be used to select healthier udders have recently been described, including electrical conductivity of milk, lactoferrin levels, cytokine

http://dx.doi.org/10.19103/AS.2019.0058.13
Published by Burleigh Dodds Science Publishing Limited, 2020.

Table 1 Potential udder health phenotypes

Type	Measure	Illustrative reference	Type	Measure	Illustrative reference
Direct	Clinical mastitis	Bramley et al. (1996)	Indirect	Changes in SCC patterns	de Haas et al. (2008)
	Subclinical mastitis	Bramley et al. (1996)		Differential SCC	Schwarz et al. (2011)
	Pathogen-specific mastitis	de Haas et al. (2004)		Electrical conductivity	Norberg et al. (2004)
Indirect	SCC	Schukken et al. (2003)		Lactoferrin content	Soyeurt et al. (2012)
	Milking speed/ milkability	Sewalem et al. (2011)		Pathogen-specific mastitis	Schukken et al. (1997)
	Udder conformation	Nash et al. (2002)		Immune response	Thompson-Crispi et al. (2012)
	Thermal imaging	Hovinen et al. (2008)			

References provided are for illustrative purposes as the literature on some traits is quite extensive.

concentrations and mid-infrared spectra of milk samples (Table 1). These phenotypes fall into two classes: direct observations of clinical or subclinical mastitis, and indirect observations of animal performance or milk composition. Indirect measurements are often more affordable and have the potential to generate lots of phenotypes in an automated fashion, but there is some imprecision because they can be affected by factors other than mastitis. These new data, in combination with existing recorded phenotypes, can be used to improve the genetic merit of milking cows, regardless of breed, for resistance to clinical mastitis. Such improvement will benefit cows, farmers and consumers.

2 Conventional phenotypes for improving resistance to clinical mastitis

Udder health improvement schemes require recording of direct or indirect indicators of mastitis. Directly recorded mastitis is, for example, the number of cases of clinical mastitis per cow per lactation. Subclinical mastitis is typically recorded using SCC as a proxy. Other traits for indirectly recording mastitis include milkability (milking speed) and udder conformation traits (e.g. udder depth, fore udder attachment teat length).

Selection for improved (decreased) somatic cell score (SCS) has been effective for the US Holstein population (Fig. 1), with both cows and bulls showing improvement in average breeding value. Similar trends have been observed in other populations, such as Norwegian Red cattle (Heringstad et al., 2007). Results from the International Bull Evaluation Service (Uppsala, Sweden) evaluations for Holstein cattle confirm earlier results

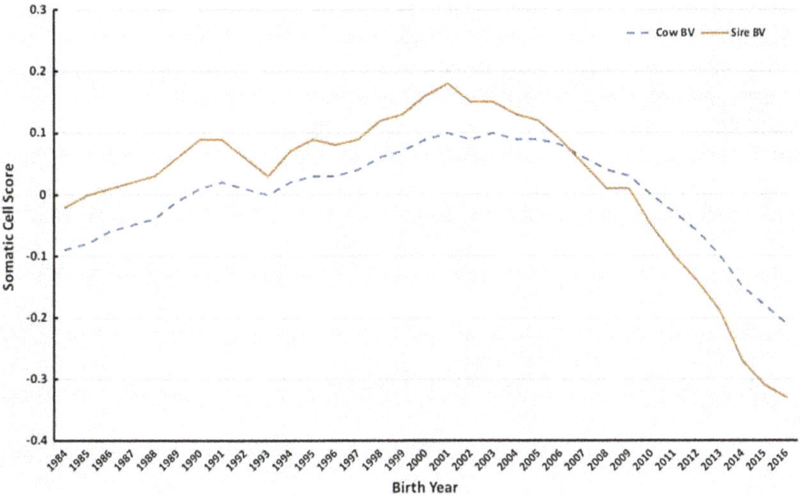

Figure 1 Genetic trend for somatic cell score [$\log_2$(somatic cell count)] in US Holsteins. Data source: Council on Dairy Cattle Breeding, Bowie, Maryland, USA; https://queries.uscdcb.com/eval/summary/trend.cfm?R_Menu=HO.s#StartBody.

showing that lower breeding values for SCC are accompanied by lower rates of clinical mastitis (Mark et al., 2002).

In many countries, reliable recording of clinical mastitis incidents is hard to achieve, so udder health improvement must begin with a focus on indirect measures. SCC is genetically correlated with clinical mastitis (r_g = 0.60–0.70). This means that when analysing field data, an observed high level of SCC is generally accompanied by a clinical or subclinical mastitis event. While milk of healthy cows commonly shows day-to-day variation in SCC, but most of the variation in SCC is associated by clinical or subclinical mastitis.

There also is some evidence that udder health decreases when herds move from conventional to automated milking (Hovinen and Pyörälä, 2011). Some of these changes may be attributable to limitations in the technology available, but others are due to trade-offs related to more frequent milking, such as reduced time for teat canal closure and attendant risk of bacterial infiltration between milkings. As herd sizes continue to increase in many countries, cows also may receive less individual treatment, which may result in less-frequent treatment for mastitis.

2.1 Clinical mastitis

Clinical mastitis is an outer visual or perceptible sign of an inflammatory response of the udder: painful, red, swollen udder. The inflammatory response can also be recognized by abnormal milk, or a general illness of the cow, with fever. Subclinical mastitis is also an inflammatory response but without outer visual or perceptible signs of the udder. An incident of subclinical mastitis is detectable with indicators such as the electrical conductivity of the milk, *N*-acetyl-β-D-glucosaminidase and cytokine concentrations and SCCs in the milk. Recording of clinical mastitis cases may be used in many ways, including veterinary support of farm management (i.e. identification of diseased animals and establishment of consistent treatment procedures), national veterinary policy-making (i.e. drug regulations and preventive epidemiological measures), addressing citizens' and consumers' concerns about animal health and welfare and product quality and safety (i.e. food chain management and product labelling) and genetic improvement (i.e. monitoring the genetic level of the population and selection and mating strategies).

Clinical mastitis ideally should be evaluated as a binary trait using a threshold model (e.g. Zwald et al., 2004; Koeck et al., 2010; Gaddis et al., 2014; Vukasinovic et al., 2017), although some genetic evaluation centres currently use linear models for health disorders (Council on Dairy Cattle Breeding, 2018). A binary trait is coded using only two categories, which would represent presence or absence of infection in the case of clinical mastitis. These traits differ from 'classical' traits, such as milk yield, because there are only two distinct values represented in the population rather than a wide range of values that commonly follows a normal distribution. From a theoretical point of view, the two types of coding should be modelled using different statistical approaches, but for purposes of ranking animals, the models are reasonably robust when assumptions are violated. Literature estimates of heritabilities for clinical mastitis range from 0.06 to 0.10 on the underlying scale (Zwald et al., 2004; Gaddis et al., 2014) and from 0.01 to 0.14 on the observed scale (e.g. Rupp and Boichard, 1999; Nash et al., 2000; Heringstad et al., 2001). Larger values, ranging from 0.21 to 0.42, have been reported from linear models (Pryce et al., 1997; Nash et al., 2000) but may be attributable in part to small datasets. When feasible, animal models are preferred to sire or sire–maternal grandsire models because they provide estimates of cow breeding values, as well as those of bulls.

2.2 Milk somatic cell count

Somatic cells in milk are primarily leukocytes or white blood cells but also include sloughed epithelial (milk-secreting) cells. Epithelial cells are always present in milk at low levels as a result of the replacement of old with new cells, with normal milk SCC levels being lower than 50000. White blood cells are present in milk in response to tissue damage and/or clinical and subclinical infections. As the degree of damage or the severity of infection increases, so does the level of white blood cells. Reduced SCC is associated with lower incidence and fewer clinical episodes of clinical mastitis; greater quality and shelf life of dairy products; increased cheese yield; and higher premium payments for milk quality. Hadrich et al. (2018) found that persistent SCC above 100000 cells/mL results in lost milk yield ranging in value from US$1.20/cow/day in the first month of lactation to US$2.06/cow/day in the tenth month of lactation.

The International Dairy Federation (IDF, 2013) provides a comprehensive set of guidelines for the measurement and interpretation of milk SCC. Thresholds for declaring that a cow is likely to have mastitis based on quarter- or cow-level SCC based on those guidelines are presented in Table 2. A bulk tank SCC threshold also is provided, and herds exceeding that limit could lose their ability to market their milk. However, it is important to note that these thresholds are not absolute indicators of infection, and animals exceeding these limits should be interpreted as having a higher risk of mastitis.

Can SCC be reduced to the point that cows are at increased risk of infection? This question was hotly debated when genetic evaluations for SCC were proposed because SCC is associated with innate immune responses to infection (Wellnitz and Bruckmaier,

Table 2 Recommended thresholds for quarter, cow and bulk tank somatic cell counts likely to indicate the presence of clinical mastitis

Level of measurement	Threshold (cells/mL)	Interpretation	Source
Quarter (teat) level	100000	Above this level, a quarter is likely to be infected	IDF (2013)
Cow level	200000	Above this level, a cow is likely to have an infected mammary gland	IDF (2013)
Bulk tank level	400000	A 3-month geometric mean bulk tank SCC above this level should be placed on a watch list and monitored	Hillerton and Berry (2004)

2012), and it does appear that some herds with very low average SCC may have reduced ability to respond to clinical infections (e.g. Suriyasathaporn et al., 2000; Beaudeau et al., 2002), although the precise meaning of 'very low' varies from study to study. An average below 50000 cells/mL is often considered undesirable by mastitis experts. This can be managed in a genetic programme by using selection index theory in a couple of different ways. First, if the average SCC in a population has reached a desirable level, then restricted selection index (Kempthorne and Nordskog, 1959) can be used to maintain that genetic level. Another alternative is to use a non-linear selection index to assign higher weight to animals with breeding values near the optimum (Thompson, 1980), which is intermediate between very high and very low SCC. It is important to emphasize that the heritability of SCS, a $\log_2$ transformation of SCC used in the United States for genetic evaluation, has a heritability of only 12%, so most variation from animal to animal is due to management and other non-genetic factors.

2.3 Milking speed (milkability)

Strictly speaking, milking speed is not a measure of udder health. Rather, it measures a physical property of the udder – how fast the milk flows from each quarter into the milking unit – that may be associated with udder health. It is of interest to this discussion because milking speed data are routinely collected by many milking systems and stored in on-farm computer systems. Genetic correlations of SCS with milking speed generally are moderate and antagonistic (e.g. Zhang et al., 1994; Boettcher et al., 1998; Rupp and Boichard, 1999), which suggests that the optimal milking speed may have an intermediate optimum. Cows that milk too quickly may have elevated risk of intramammary infections, while cows that milk very slowly disrupt milk procedures. The latter case is of growing concern as more farms switch to robotic milking, where there is a need to minimize the number of milking unit purchases while ensuring that robots are available when cows want to be milked.

Milking speed has appeal as a correlated trait because, on many farms, the cost of data collection is minimal. However, there is no consistent scale used for milking speed across models within a manufacturer, or across manufacturers. Even systems which record actual (wall clock) milking times may produce records that are not generally comparable because of differences in when during the milking process recording begins and ends. Some national genetic evaluations for milkability are based on qualitative (Wiggans et al., 2007), rather than quantitative, scales. Such scales generally express milking speed in discrete categories ranging from 'much slower than average' to 'much faster than average'.

2.4 Udder conformation

Linear scoring of udder conformation is recommended by the World Holstein Friesian Federation (WHFF[1]) and International Committee for Animal Recording (ICAR[2]). A full description of conformation traits is given in Section 05 of the ICAR Guidelines, as well as in the WHFF report 'Progress of type harmonisation, May 2016',[3] and traits should be scored according to those recommendations.

Genetic correlations of udder depth and fore udder attachment with SCS and clinical mastitis suggest that these traits should be included in selection indices to help improve udder health. Some teat conformation traits (e.g. Seykora and McDaniel, 1985), which are not routinely scored, have also been associated with the ability of the mammary gland to resist infection by preventing the infiltration of microorganisms through the teat canal. It is possible that some of these traits may be routinely recorded in the future because teat conformation is important in the context of automated (robotic) milking systems (Jacobs and Siegford, 2012).

3 New phenotypes for improving resistance to clinical mastitis

Several of the phenotypes described in this section are not 'new' in the sense of being recently discovered, but advances in technology now make their direct or indirect measurement on a large scale feasible, when it previously was not. Ideally, new traits will have low genetic and phenotypic correlations with existing traits (there is a lot of additional information in the new observations), or the cost of recording will be very low so that many new phenotypes can be collected rapidly to provide high-reliability predicted transmitting ability (PTA).

3.1 Electrical conductivity of milk

Electrical conductivity is measured by most modern milking systems, and milk produced by cows with mastitis has higher conductivity than milk from healthy animals because of increased Na^+ and Cl^- levels (Norberg et al., 2004). Conductivity measurements at milking can also be compared with previous measurements to identify changes consistent with subclinical mastitis. However, studies show substantial variation in both sensitivity (true positive rate) and specificity (true negative rate), although modern milking systems that take measurements at the quarter level produce better results than systems that pooled milk samples. Norberg et al. (2004) showed that simple thresholds can be used to differentiate between healthy, subclinical and clinical cows with reasonable sensitivity and specificity. For example, a threshold of 1.15 applied to the inter-quarter ratio between the maximum and minimum averages of the 20 highest valid electrical conductivity measures

1. http://www.whff.info/documentation/typeharmonisation.php#go1
2. http://www.icar.org/Guidelines/05-Conformation-Recording.pdf
3. http://www.whff.info/documentation/documents/progressoftypeharmonisationversionafterBuenosAiresv2.pdf

taken during a milking correctly classified 80.6% of clinically and 45.0% of subclinically infected cows, as well as 74.8% of the healthy cows. More sophisticated models may provide higher sensitivity and specificity at the cost of greater complexity (Norberg, 2005).

3.2 Lactoferrin

Lactoferrin is an iron-binding glycoprotein naturally present in milk that is a major component of the mammalian innate immune system (González-Chávez et al., 2009), and it also is released by neutrophils during inflammation. Elevated lactoferrin levels are, therefore, indicative of a physiological response to infection and may be used to diagnose clinical mastitis (Shimazaki and Kawai, 2017). Soyeurt et al. (2012) showed that MIR spectroscopy can be used to cheaply and rapidly predict milk lactoferrin content, which may be useful as an indirect indicator of mastitis. Lactoferrin is significantly higher in cows with clinical mastitis than those without, and there also appear to be differences between animals with environmental and infectious mastitis (Kawai et al., 1999). Healthy cows averaged 169 μg/mL of lactoferrin, cows with subclinical mastitis averaged 495 μg/mL, and animals with clinical mastitis averaged 895 μg/mL. Differences in lactoferrin concentration were significant for each of these groups. Soyeurt et al. (2012) proposed a threshold of 200 μg/mL of lactoferrin to differentiate between healthy and sick animals.

3.3 N-acetyl- -D-glucosaminidase

N-acetyl-β-D-glucosaminidase (NAGase) is a lysosomal enzyme that is released into milk from neutrophils during phagocytosis and cell lysis, as well as from damaged epithelial cells (Pyörälä, 2003). Hovinen et al. (2016) reported that NAGase activity can be used to detect both subclinical and clinical mastitis with high levels of accuracy, although Nyman et al. (2016) reported that SCC was the best overall predictor of intra-mammary infection. If the cost of recording NAGase levels is competitive with that of SCC and can easily fit into a milk-testing laboratory's workflow, then there may be value in routinely recording the phenotype.

3.4 Pathogen-specific mastitis

Pathogens associated with contagious mastitis (e.g. *Staphylococcus aureus*) produce different patterns of SCC than do pathogens associated with environmental mastitis (e.g. *Escherichia coli*, *Streptococcus uberis*). This is because different pathogens stimulate responses by different parts of the immune system (innate versus adaptive responses; e.g. Schukken et al., 1997). Bacteriological cultures are not routinely used to identify the causative organism for cases of clinical mastitis, and the cost of doing so is likely to prevent such data from being available for routine genetic evaluation. Patterns of infection differ among pathogens, with some species (e.g. *E. coli*) being primarily responsible for clinical infections, while others (e.g. *S. aureus*) are primarily responsible for subclinical infections (Schukken et al., 1997). Knowledge of the causative pathogen could provide useful information for modelling clinical and subclinical mastitis with greater precision. Pathogen information from the mastitis laboratories are recorded routinely in Norway (Haugaard et al., 2012), Denmark (Sørensen et al., 2009), Sweden (Holmberg et al., 2012) and Finland (Koivula et al., 2007). This information can be combined with other relevant information, such as CM or SCC, to define pathogen-specific mastitis for individual cows.

3.5 Patterns of somatic cells in milk

De Haas et al. (2008) compared several traits computed from test-day SCC and patterns of peaks in SCC against lactation-average SCC for their ability to detect clinical mastitis. The heritabilities of the new traits ranged from 0.01 to 0.11, and genetic correlations with clinical and subclinical mastitis ranged from 0.60 to 0.93 and 0.55 to 0.98, respectively. Different patterns of SCC are associated with different pathogens, and adding that information to models could improve prediction accuracy. However, monthly intervals are too long to capture changes in SCC due to organisms such as *E. coli* that cause rapid, acute infections. An important point noted by de Haas et al. (2008) and others (e.g. Schepers et al., 1997) is that log transformation of SCC to produce an SCS tends to reduce high test-day SCC. More recently, Bobbo et al. (2018) found that novel traits derived from SCC had heritabilities at least as large as SCS but were sensitive to environmental effects. New traits may be useful for improving current over previous lactation udder health, but care is needed to ensure that models properly account for environmental factors. Test-day SCC records should be stored in addition to SCS records so that patterns in SCC can be analysed, as well as lactation average SCC.

3.6 Differential somatic cell count

The distribution of leukocytes (white blood cells) is different in milk from healthy and infected mammary glands (Nickerson, 1989). Differential somatic cell counts (DSCCs) are used to quantify the proportions of different types of cells in the mammary gland and may be used to identify cases of subclinical mastitis that cannot be detected by SCC alone (Pilla et al., 2013). Patterns of DSCC also may be used to distinguish between acute and chronic mastitis (Leitner et al., 2000). Piepers et al. (2009) described a method for the flow cytometric quantification of the proportion of viable, apoptotic and necrotic polymorphonuclear neutrophilic leukocytes in cow's milk that can serve as the basis of automated, routine phenotype collection. Subsequent research has also shown that DSCC can be used to identify inflammatory responses in quarters that are classified as healthy based on overall SCC (Schwarz et al., 2011). Damm et al. (2017) recently demonstrated that DSCC and SCC can be reliably and repeatably estimated simultaneously, at low cost, in commercial milk-testing laboratories using a method developed by Foss Analytical A/S (Hilleroed, Denmark; Holm, 2013). The principal obstacle to the adoption of routine DSCC now appears to be the availability of suitable equipment in testing laboratories. The Fossomatic 7 DC and CombiFoss 7 DC instruments now support routine collection of DSCC, and it is anticipated that other manufacturers will develop their own products that will provide similar analyses.

Schwarz (2017) suggested that a DSCC threshold of 75% could be used to distinguish between active and inactive inflammatory responses but noted that a two-factor classification system involving both SCC and DSCC is more useful for categorizing test-day milk samples. Such a scheme is described in Table 3 and allows users to differentiate among clinical and chronic infections, as well as subclinical cases of mastitis.

3.7 Thermal imagery

Berry et al. (2003) showed that infrared thermography (IRT) could be used to predict actual udder surface temperatures and proposed that thermal imagery could be used as

Table 3 Decision support grid for classifying mastitis status based on somatic cell count (SCC) and differential somatic cell count (DSCC) measurements

	Low DSCC (≤75%)	High DSCC (>75%)
Low SCC (≤100 000 cells/mL)	Normal/healthy mammary gland	Onset/early stage of clinical mastitis (SCC <100 000 cells/mL and elevated proportions of PMN)
High SCC (>100 000 cells/mL)	Chronically infected cows	The cow's immune system is actively fighting mastitis pathogens

a predictor of inflammation associated with mastitis. This is appealing because the cost of thermal imagers has decreased steadily over the last several years, which enables their use in precision dairy settings. Colak et al. (2008) reported that IRT can be used to differentiate between udder surface temperatures that are associated with varying degrees of infection, and subsequent studies (e.g. Hovinen et al., 2008; Bortolami et al., 2015) supported these findings but found that IRT data are not useful for identifying causal organisms of infection. If the cost of installing thermal cameras is sufficiently low, changes in udder skin surface temperature identified using IRT may be a useful mastitis indicator, particularly when combined with other data on animal behaviour and milk composition.

3.8 Collection of new phenotypes

The expected benefit of developing large reference populations for new phenotypes is unclear and will be closely associated with the cost of data collection. The most promising phenotypes may be indirect predictors developed from mid-infrared spectral data because many milk-testing laboratories are now equipped with instruments for that analysis. However, there is still infrastructure needed for calibration, data collection and data transfer before that information can routinely feed into genetic improvement programmes. If the cost of on-farm sensors continues to decrease rapidly, we may see a substantial increase in per-milking SCC data collected before spectral data gain ground. Regardless of the technology, there has to be a clear benefit to farmers from new data if they are being asked to pay for it. The best way to collect phenotypes for new traits may be to pay for extensive phenotyping in a small group of herds with high-quality data (e.g. Chesnais et al., 2016; Schöpke and Swalve, 2016), but the discussion remains largely theoretical at this time.

4 National and international genetic improvement programmes for resistance to clinical mastitis

4.1 International evaluations

The International Bull Evaluation Service (Interbull, Uppsala, Sweden) distributes genetic evaluations for SCC and clinical mastitis for Holsteins in member countries. Twenty-seven countries (Australia, Belgium, Canada, Croatia, Czech Republic, Denmark & Sweden & Finland, Estonia, France, Germany & Austria & Luxembourg, Great Britain, Hungary,

Ireland, Israel, Italy, Japan, Latvia, Lithuania, the Netherlands (including the Belgium Flemish region), New Zealand, Poland, Portugal, Republic of Korea, Republic of South Africa, Slovenia, Slovak Republic, Spain and Switzerland) currently participate in the Interbull SCC evaluations, and five (Canada, Denmark & Sweden & Finland, France, Great Britain, the Netherlands (including the Belgium Flemish region)) in the clinical mastitis evaluations. The United States has submitted data for the August 2018 test run but is not yet an official participant in the clinical mastitis evaluations. International evaluations are not available for other breeds due to the limited number of daughter records available for health traits in non-Holstein populations.

4.2 Total merit indices

Many countries use total merit indices (TMI) as their national selection objective. Selection indices (e.g. Cole and VanRaden, 2018) combine information about many traits into a single criterion that can be used for ranking and selecting animals. Indices differ from one country to another because economic conditions, farm policies and markets differ. Figure 2 shows the weights assigned to different traits and trait groups in the TMI of 21 different countries. While there are similarities between indices, there are also substantial differences. For example, the US lifetime net merit index (NM$; Cole and VanRaden, 2018) has almost no weight on mild yield, but the US fluid merit index includes substantial weight on milk. These differences reflect differential payment for milk components in different parts of the US. Every index shown includes selection for improved udder health, either indirectly through SCS or directly for udder health traits.

Changes in rates of genetic change after adding new traits to a selection index may be more limited than initially assumed because of correlations among traits. Table 4 shows correlations of PTA for six health traits in US Holsteins with several production, fertility and fitness traits. Absolute correlations with traits already evaluated range from lows near 0.15 to a high near 0.70. Notably, correlations with longevity (productive life) and fertility (daughter pregnancy rate) are fairly high, which means that there has been selection for improved disease resistance in the US Holstein population for many years despite the lack of direct measures of health for most of that time. Clinical mastitis also has a significant correlation with SCS, which has been in the NM$ index since 1994, and the genetic trend for clinical mastitis shows a favourable trend (Fig. 3) that is due, in part, to correlated response to selection for reduced SCS. The 2017 version of NM$, which did not include the six direct health traits, had a correlation of 0.47 with the PTA for HTH$, which is the lifetime value of all health costs for an individual (VanRaden, Cole and Parker Gaddis, 2018). This is virtually identical to the correlation of the 2018 revision, which does include the health trait, with an HTH$ of 0.46.

4.3 Effects of indicator traits in selection indices

The accuracy of a selection index is based on the genetic parameters of the individual traits in the index, as well as the phenotypic and genetic correlations of the individual traits with one another. When heritabilities and/or correlations of new with existing traits are low, then the reliabilities may be reduced, even if the new traits are more related biologically to the true phenotype of interest. Less-precise traits with large numbers of existing observations may also produce higher reliabilities than more-precise traits with few observations. These issues have been explored in detail by Gonzalez-Recio et al. (2014).

Published by Burleigh Dodds Science Publishing Limited, 2020.

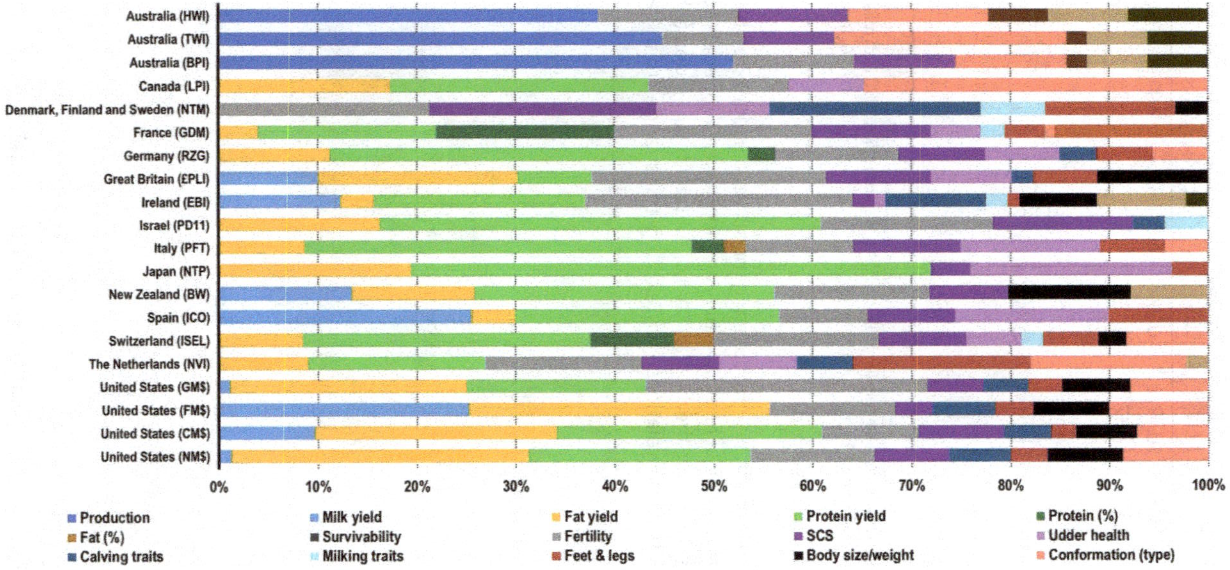

Figure 2 Traits included in 21 total merit indices of the United States and 16 other countries. Data were collected from genetic evaluation centres and purebred cattle associations for Australia (ADHIS, 2014); Canada (CDN, 2017); Denmark, Finland and Sweden (NAV, 2017); France (Genes Diffusion, 2014); Germany (VIT, 2017); Great Britain (AHDB Dairy, 2017); Ireland (ICBF, 2017); Israel (SION, 2015); Italy (ANAFI, 2016); Japan (Holstein Cattle Association of Japan, 2010); New Zealand (DairyNZ, 2017); Spain (CONAFE, 2016); Switzerland (Holstein Association of Switzerland, 2013); the Netherlands (CRV, 2017); and the United States (Holstein Association USA Inc., 2017; VanRaden, 2017). Index abbreviations are HWI = health weighted index; TWI = type weighted index; BPI = balanced performance index; LPI = lifetime profit index; NTM = Nordic total merit; GDM = genes diffusion merit; RZG = Relativ Zuchtwert Gesamt (total merit index); £PLI = profitable lifetime index; EBI = economic breeding index; PD11 = Israeli 2011 breeding index; PFT = production, functionality and type index; NTP = Nippon total profit; BW = breeding worth; ICO = Índice de Mérito Genético Total (total genetic merit index); ISEL = Index de Sélection Totale (total selection index); NVI = Netherlands cattle improvement index; TPI = total performance index; GM$ = grazing merit; FM$ = fluid merit; CM$ = cheese merit; NM$ = net merit. Source: after Fig. 4 in Cole and VanRaden (2018).

Table 4 Correlations of six producer-recorded health traits in US Holsteins with protein yield (PRO), productive life (PL), cow livability (LIV), somatic cell score (SCS), daughter pregnancy rate (DPR), cow conception rate (CCR) and heifer conception rate (HCR)

Health trait	PRO	PL	LIV	SCS	DPR	CCR	HCR
Hypocalcaemia	0.18	0.15	0.19	*–0.29*	0.003	0.01	0.02
Displaced abomasum	0.23	*0.35*	*0.47*	–0.13	*0.32*	*0.28*	0.24
Ketosis	0.03	0.33	0.27	–0.19	*0.59*	*0.49*	0.07
Mastitis	0.06	*0.39*	*0.22*	*–0.68*	*0.20*	*0.21*	0.06
Metritis	0.05	*0.32*	*0.26*	–0.09	*0.46*	*0.41*	*0.23*
Retained placenta	–0.03	*0.17*	*0.13*	–0.10	*0.14*	*0.13*	*0.12*

Correlations for hypocalcaemia and retained placenta were calculated using PTA bulls born since 1990 with reliability of 75%. All other correlations were calculated using PTA for bulls born since 1990 with reliability ≥90%. Italicized correlations are different from 0.
Source: Council on Dairy Cattle Breeding, Bowie, Maryland, USA.

For example, the United States recently introduced genomic evaluations for health traits to complement existing evaluations of health, fertility and longevity (VanRaden et al., 2018). The heritability of SCS is 12%, but the heritability of the health sub-index (which includes CM) is only 1%. Phenotypic correlations with existing traits range from 1% to

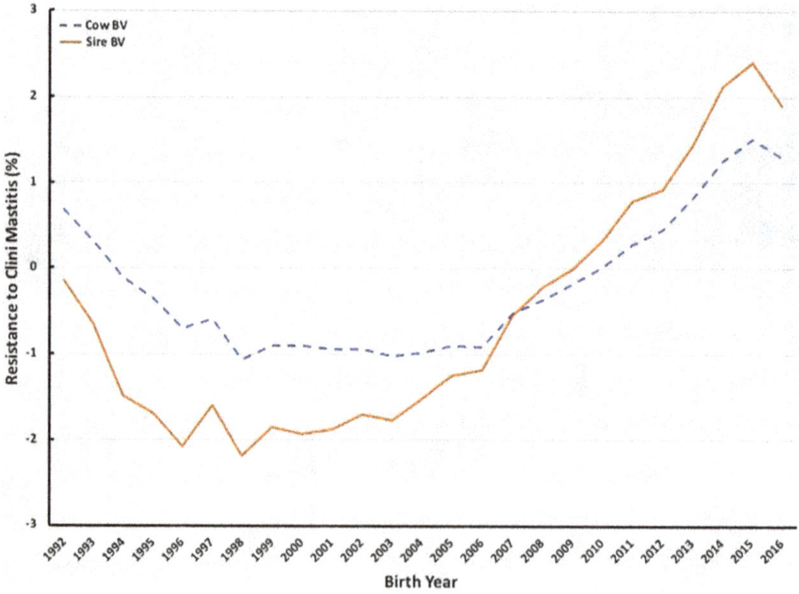

Figure 3 Genetic trend for resistance to clinical mastitis in US Holsteins. Data source: Council on Dairy Cattle Breeding, Bowie, Maryland, USA.

28%, and genetic correlations with existing traits range from 1% to 56%. However, there are approximately 2 million CM records on 1.1 million Holstein cows, and there are more than 56 million SCS records from 23 million Holstein cows. If the less-precise trait (SCS) was replaced with the more-precise trait, the reliability of the resulting index values would be lower and genetic progress would be reduced relative to an index that includes both.

5 Increasing rates of genetic gain through genomic selection

The rapid adoption of genomic selection by all of the major dairy-producing countries (e.g. Wiggans et al., 2017) has resulted in dramatic decreases in selection intervals and increases in rates of genetic gain (García-Ruiz et al., 2016), and has also supported the development of national evaluations for low-heritability traits with a limited number of phenotypes available (e.g. Pryce and Daetwyler, 2012; Gaddis et al., 2014; Chesnais et al., 2016). The success of this approach is demonstrated in Fig. 4, which shows selection differentials of SCS in US Holsteins for the four paths of selection (lower SCS is desirable, so negative selection differentials are favourable). There is a substantial increase in the rate of change of selection differentials following the launch of genomic evaluations in 2009. As a result of this new selection tool, genetic merit for resistance to mastitis and SCS of sires of bulls and cows and dams of bulls rapidly increased in the US (Figs. 1 and 3).

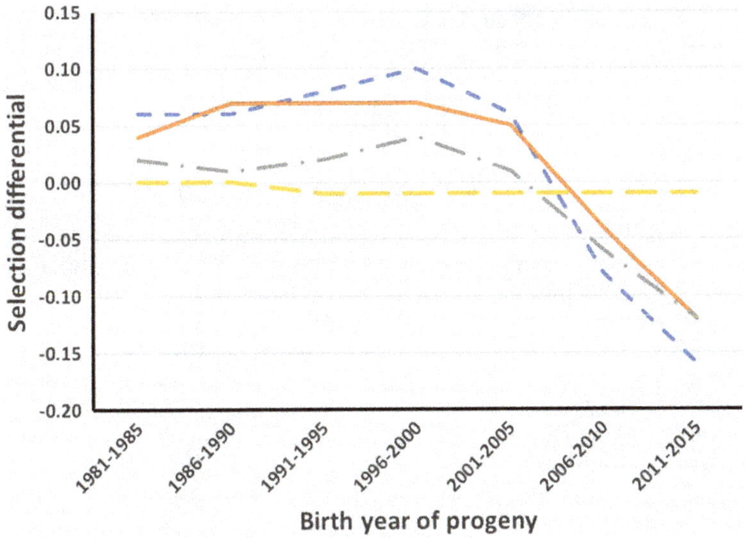

Figure 4 Selection differentials for somatic cell score of the sires of bulls (broken blue line), sires of cows (solid orange line), dams of bulls (dotted-and-dashed grey line) and dams of cows (long dashed yellow line) paths in US Holsteins for 5-year windows between 1981 and 2015. Source: figure created by author from values in Table S2 of García-Ruiz et al. (2016).

Table 5 Mean reliability (%) of traditional and genomic evaluations of young and progeny-tested bulls for six producer-recorded health traits of US Holsteins

	Progeny-tested bulls[a]			Young bulls[b]		
Health trait	Traditional	Genomic	Gain	Traditional	Genomic	Gain
Hypocalcemia	20.0	44.2	24.2	10.9	40.0	29.1
Displaced abomasum	25.7	47.1	21.4	14.6	41.8	27.2
Ketosis	24.0	46.2	22.2	13.4	41.2	27.8
Mastitis	33.3	56.3	23.0	18.3	49.4	31.1
Metritis	27.6	48.1	20.5	15.4	42.2	26.8
Retained placenta	25.6	46.7	21.1	14.2	41.6	27.4

[a] Progeny-tested bulls proofs include daughter information.
[b] Young bull proofs include only parent average and genomic information.
Source: Council on Dairy Cattle Breeding, Bowie, Maryland, USA.

Some countries, such as Norway, have been collecting health data for decades (e.g. Heringstad and Østerås, 2013) and can compute breeding values with reasonable reliabilities for proven bulls. In other countries, data have been available for a much shorter amount of time and, without genomics, reliabilities for most bulls are too low for publication. Table 5 shows the reliability gains for six health traits for US Holstein cattle. The average reliability of clinical mastitis, which has a heritability of 0.03 in that population, increased by 23.0 in proven bulls and 31.1 in young bulls, resulting in average genomic reliabilities of 56.3 and 49.4. These values are similar to those of longevity and fertility traits that are routinely evaluated in the US (VanRaden et al., 2009). This provides an opportunity for countries with shorter histories of data collection to compute useable evaluations for health and fitness traits, such as resistance to clinical mastitis.

5.1 Opportunities for marker-assisted selection

If causal DNA variants with large effects on individual traits can be identified, then marker-assisted selection can be used to increase the frequency of these desirable alleles (e.g. Dentine, 1992). Quantitative trait loci (QTL) associated with udder health have been mapped to many regions of the genome, including chromosomes 6, 11, 13, 14, 18, 20, 24 and 29 in various Holstein populations (e.g. Ashwell et al., 1996; Schrooten et al., 2000; Klungland et al., 2001; Kuhn et al., 2003; Sahana et al., 2013; Tiezzi et al., 2015).

The Animal QTL Database (AnimalQTLdb[4]) provides an exhaustive list of putative QTL associated with CM, SCC and SCS collected from the literature. The large number of QTL identified precludes an exhaustive discussion of them all, but Fig. 5 shows the number of entries in AnimalQTLdb for three udder health traits. While it appears that there are, in general, many more QTL associated with SCS than CM or SCC, that is because there are so many studies on SCS. Each QTL region identified is reported in AnimalQTLdb, which inflates the counts.

4 https://www.animalgenome.org/cgi-bin/QTLdb/index

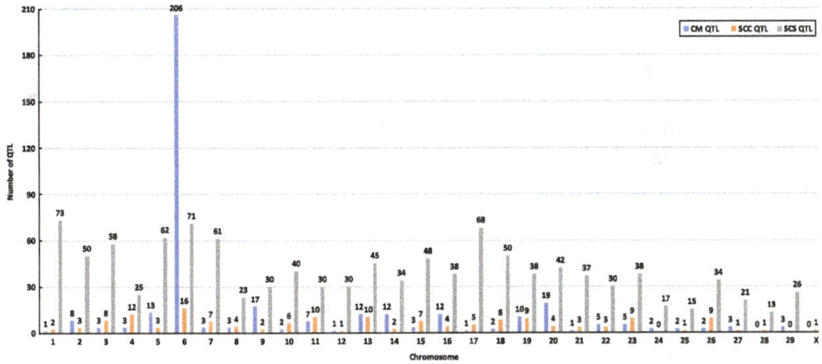

Figure 5 The number of quantitative trait loci (QTL) in the Animal QTL Database on each bovine chromosome for clinical mastitis (CM), somatic cell count (SCC) and somatic cell score (SCS). Note that these results include all breeds represented in the database, and QTL may overlap across studies. Source: figure created by author using data retrieved from AnimalQTLdb: https://www.animalgenome.org/tmp/map554722875.txt.gz, https://www.animalgenome.org/tmp/map121076229.txt.gz and https://www.animalgenome.org/tmp/map185863377.txt.gz.

6 Conclusion

The major gap to be bridged in order to produce cows that are more genetically resistant to clinical mastitis is that between research and production. The phenotypes most commonly used in genetic improvement programmes are those that are the easiest to measure in many cows, such as SCC. However, those traits will not result in the highest rates of genetic gain because they share only some of the same biological mechanisms in common. While researchers continue to identify more precise measurements of individual infection status, the cost of phenotyping often is high, and many require the purchase of specialized equipment.

The focus of this chapter has been on genetic improvement, but the role of herd management should not be overlooked. As the low heritabilities of different measures of clinical mastitis attest, most of the variation among individuals and between farms is attributable to environmental differences. Management practices that minimize the ability of pathogens to survive on the farm, and which limit transmission from animal to animal, should be identified and promoted in conjunction with genetic improvement programmes.

7 Future trends in research

International efforts such as the Functional Annotation of Animal Genomes project (Andersson et al., 2015) are working to identify true variants that can be used as targets for genomic selection and gene editing. The identification of key regulatory elements also may provide new therapeutic targets that can be used to improve mastitis resistance. As briefly discussed above, research also continues on new technologies that can be used to rapidly phenotype many individuals. Those phenotypes will support both improved on-farm decision-making as well as genetic improvement programmes.

There is tremendous interest in the use of new molecular biology tools, such as clustered regularly interspaced short palindromic repeats (i.e. CRISPR-Cas9), to make precise, highly targeted changes to animals' genomes. Such approaches may be more acceptable to consumers and regulatory agencies than previous transgenic approaches (e.g. Wall et al., 2005). Some gene-edited products have recently reached the US marketplace (Ledford, 2015; Waltz, 2016), but considerable uncertainty remains about the manner in which gene-edited plant and animal products will be regulated (Maxmen, 2017).

8 Acknowledgements

The author was supported by the appropriated project 8042-31000-002-00-D, 'Improving Dairy Animals by Increasing Accuracy of Genomic Prediction, Evaluating New Traits, and Redefining Selection Goals', of the Agricultural Research Service of the United States Department of Agriculture (USDA-ARS). Mention of trade names or commercial products in this chapter is solely for the purpose of providing specific information and does not imply recommendation or endorsement by the USDA. The USDA is an equal opportunity provider and employer.

9 Where to look for further information

9.1 Introductory works

'Current Concepts of Bovine Mastitis', published by the National Mastitis Council,[5] provides an excellent overview of the topic of mastitis. This is a seminal work on the subject that is recommended to anyone interested in udder health.

9.2 Key societies

Important international organizations include the American Dairy Science Association,[6] International Committee for Animal Recording,[7] International Dairy Federation[8] and National Mastitis Council.[9] The International Bull Evaluation Service[10] focusses strictly on genetic evaluation practices, which are critical for long-term improvements to mastitis resistance. These organizations publish scientific journals, organize meetings and promote international standards to improve animal health and milk quality.

5 https://www.nmconline.org/publications/
6 https://www.adsa.org/
7 https://www.icar.org/
8 https://www.fil-idf.org/
9 https://www.nmconline.org/
10 http://www.interbull.org/index

9.3 Key journals and conferences

The *Journal of Dairy Science*[11] is the most prominent scientific publication in the field, and there are frequently multiple sessions on mastitis at the American Dairy Science Association Annual Meeting. The National Mastitis Council holds an annual meeting specifically on mastitis and related topics, and the *National Mastitis Council Annual Proceedings*[12] commonly publishes reports on cutting-edge research before its publication in peer-reviewed journals. Interbull publishes the *Interbull Bulletin*[13] and hosts annual meetings and workshops, which often include technical reports on genetic evaluation methodology for traits related to mastitis resistance.

9.4 Other resources

The Animal QTL Database (AnimalQTLdb[14]) provides extensive information on putative QTL related to clinical mastitis, SCC and SCS assembled from more than 800 scientific publications. This is an excellent resource to identify genomic regions that have large effects on udder health traits, and it is frequently updated.

10 References

Andersson, L., Archibald, A. L., Bottema, C. D., Brauning, R., Burgess, S. C., Burt, D. W., Casas, E., Cheng, H. H., Clarke, L., Couldrey, C., et al. 2015. Coordinated international action to accelerate genome-to-phenome with FAANG, the Functional Annotation of Animal Genomes project. *Genome Biology* 16(1), 57. doi:10.1186/s13059-015-0622-4.

Ashwell, M. S., Rexroad, C. E., Miller, R. H. and VanRaden, P. M. 1996. Mapping economic trait loci for somatic cell score in Holstein cattle using microsatellite markers and selective genotyping. *Animal Genetics* 27(4), 235–42. doi:10.1111/j.1365-2052.1996.tb00484.x.

Beaudeau, F., Fourichon, C., Seegers, H. and Bareille, N. 2002. Risk of clinical mastitis in dairy herds with a high proportion of low individual milk somatic-cell counts. *Preventive Veterinary Medicine* 53(1–2), 43–54. doi:10.1016/S0167-5877(01)00275-6.

Berry, R. J., Kennedy, A. D., Scott, S. L., Kyle, B. L. and Schaefer, A. L. 2003. Daily variation in the udder surface temperature of dairy cows measured by infrared thermography: potential for mastitis detection. *Canadian Journal of Animal Science* 83(4), 687–93. doi:10.4141/A03-012.

Bobbo, T., Penasa, M., Finocchiaro, R., Visentin, G. and Cassandro, M. 2018. Alternative somatic cell count traits exploitable in genetic selection for mastitis resistance in Italian Holsteins. *Journal of Dairy Science* 101(11), 10001–10. doi:10.3168/jds.2018-14827.

Boettcher, P. J., Dekkers, J. C. and Kolstad, B. W. 1998. Development of an udder health index for sire selection based on somatic cell score, udder conformation, and milking speed. *Journal of Dairy Science* 81(4), 1157–68. doi:10.3168/jds.S0022-0302(98)75678-4.

Bortolami, A., Fiore, E., Gianesella, M., Corrò, M., Catania, S. and Morgante, M. 2015. Evaluation of the udder health status in subclinical mastitis affected dairy cows through bacteriological culture, somatic cell count and thermographic imaging. *Polish Journal of Veterinary Sciences* 18(4), 799–805. doi:10.1515/pjvs-2015-0104.

Bramley, A. J., et al. 1996. Current concepts of bovine mastitis. In: *Proc. 37th Ann. Mtg. Natl. Mast. Counc.*, vol. 37, pp. 1–3.

11 http://journalofdairyscience.org/
12 https://www.nmconline.org/publications/
13 https://journal.interbull.org/
14 https://www.animalgenome.org/cgi-bin/QTLdb/index

Chesnais, J. P., Cooper, T. A., Wiggans, G. R., Sargolzaei, M., Pryce, J. E. and Miglior, F. 2016. Using genomics to enhance selection of novel traits in North American dairy cattle. *Journal of Dairy Science* 99(3), 2413–27. doi:10.3168/jds.2015-9970.

Colak, A., Polat, B., Okumus, Z., Kaya, M., Yanmaz, L. E. and Hayirli, A. 2008. Short communication: early detection of mastitis using infrared thermography in dairy cows. *Journal of Dairy Science* 91(11), 4244–8. doi:10.3168/jds.2008-1258.

Cole, J. B. and VanRaden, P. M. 2018. Symposium review: possibilities in an age of genomics: the future of selection indices. *Journal of Dairy Science* 101(4), 3686–701. doi:10.3168/jds.2017-13335.

Council on Dairy Cattle Breeding. 2018. Trait reference sheet: resistance to mastitis (MAST). Available at: https://www.uscdcb.com/wp-content/uploads/2018/03/CDCB-Reference-Sheet-MAST-03_2018.pdf.

Damm, M., Holm, C., Blaabjerg, M., Bro, M. N. and Schwarz, D. 2017. Differential somatic cell count—a novel method for routine mastitis screening in the frame of Dairy Herd Improvement testing programs. *Journal of Dairy Science* 100(6), 4926–40. doi:10.3168/jds.2016-12409.

Dentine, M. R. 1992 Marker-assisted selection in cattle. *Animal Biotechnology* 3(1), 81–93. doi:10.1080/10495399209525764.

Gaddis, K. L. P., Cole, J. B., Clay, J. S. and Maltecca, C. 2014. Genomic selection for producer-recorded health event data in US dairy cattle. *Journal of Dairy Science* 97(5), 3190–9. doi:10.3168/jds.2013-7543.

García-Ruiz, A., Cole, J. B., VanRaden, P. M., Wiggans, G. R., Ruiz-López, F. J. and Van Tassell, C. P. 2016. Changes in genetic selection differentials and generation intervals in US Holstein dairy cattle as a result of genomic selection. *Proceedings of the National Academy of Sciences of the United States of America* 113(28), E3995–4004. doi:10.1073/pnas.1519061113.

González-Chávez, S. A., Arévalo-Gallegos, S. and Rascón-Cruz, Q. 2009. Lactoferrin: structure, function and applications. *International Journal of Antimicrobial Agents* 33(4), 301.e1–8. doi:10.1016/j.ijantimicag.2008.07.020.

Gonzalez-Recio, O., Coffey, M. P. and Pryce, J. E. 2014. On the value of the phenotypes in the genomic era. *Journal of Dairy Science* 97(12), 7905–15. doi:10.3168/jds.2014-8125.

de Haas, Y., Veerkamp, R. F., Barkema, H. W., Gröhn, Y. T. and Schukken, Y. H. 2004. Associations between pathogen-specific cases of clinical mastitis and somatic cell count patterns. *Journal of Dairy Science* 87(1), 95–105. doi:10.3168/jds.S0022-0302(04)73146-X.

de Haas, Y., Ouweltjes, W., ten Napel, J., Windig, J. J. and de Jong, G. 2008. Alternative somatic cell count traits as mastitis indicators for genetic selection. *Journal of Dairy Science* 91(6), 2501–11. doi:10.3168/jds.2007-0459.

Hadrich, J. C., Wolf, C. A., Lombard, J. and Dolak, T. M. 2018. Estimating milk yield and value losses from increased somatic cell count on US dairy farms. *Journal of Dairy Science* 101(4), 3588–96.

Haugaard, K., Heringstad, B. and Whist, A. C. 2012. Genetic analysis of pathogen-specific clinical mastitis in Norwegian Red cows. *Journal of Dairy Science* 95(3), 1545–51. doi:10.3168/jds.2011-4522.

Heringstad, B. and Østerås, O. 2013. More than 30 years of health recording in Norway. In: Egger-Danner, C., Hansen, O. K., Stock, K., Pryce, J. E., Cole, J., Gengler, N. and Heringstad, B. (Eds), *Challenges and Benefits of Health Data Recording in the Context of Food Chain Quality, Management and Breeding*. ICAR Technical Series no. 17. Aarhus, Denmark, pp. 39–46. Available at: http://www.icar.org/wp-content/uploads/2015/09/tec_series_17_Aarhus.pdf.

Heringstad, B., Rekaya, R., Gianola, D., Klemetsdal, G. and Weigel, K. A. 2001. Bayesian analysis of liability of clinical mastitis in Norwegian cattle with a threshold model: effects of data sampling method and model specification. *Journal of Dairy Science* 84(11), 2337–46. doi:10.3168/jds.S0022-0302(01)74682-6.

Heringstad, B., Rekaya, R., Glanola, D., Klemetsdal, G. and Welgel, K. A. 2003. Genetic change for clinical mastitis in Norwegian cattle: a threshold model analysis. *Journal of Dairy Science* 86(1), 369–75. doi:10.3168/jds.S0022-0302(03)73615-7.

Heringstad, B., Klemetsdal, G. and Steine, T. 2007. Selection responses for disease resistance in two selection experiments with Norwegian Red cows. *Journal of Dairy Science* 90(5), 2419–26. doi:10.3168/jds.2006-805.

Hillerton, J. E. and Berry, E. A. 2004. Quality of the milk supply: European regulations versus practice. In: *Proc. 43rd Ann. Mtg. Natl. Mast. Counc.*, vol. 43, pp. 207–14.

Holm, C. 2013. Method for determining a degree of infection. Patent number: EP2630487.

Holmberg, M., Fikse, W. F., Andersson-Eklund, L., Artursson, K. and Lundén, A. 2012. Genetic analyses of pathogen-specific mastitis. *Journal of Animal Breeding and Genetics = Zeitschrift Fur Tierzuchtung und Zuchtungsbiologie* 129(2), 129–37. doi:10.1111/j.1439-0388.2011.00945.x.

Hovinen, M. and Pyörälä, S. 2011. Invited review: udder health of dairy cows in automatic milking. *Journal of Dairy Science* 94(2), 547–62. doi:10.3168/jds.2010-3556.

Hovinen, M., Siivonen, J., Taponen, S., Hänninen, L., Pastell, M., Aisla, A. M. and Pyörälä, S. 2008. Detection of clinical mastitis with the help of a thermal camera. *Journal of Dairy Science* 91(12), 4592–8. doi:10.3168/jds.2008-1218.

Hovinen, M., Simojoki, H., Pösö, R., Suolaniemi, J., Kalmus, P., Suojala, L. and Pyörälä, S. 2016. N-acetyl -β-D-glucosaminidase activity in cow milk as an indicator of mastitis. *The Journal of Dairy Research* 83(2), 219–27. doi:10.1017/S0022029916000224.

IDF. 2013. Bulletin of the IDF N° 466/2013: Guidelines for the use and interpretation of bovine milk somatic cell counts (SCC) in the dairy industry. International Dairy Federation, Brussels, Belgium.

Jacobs, J. A. and Siegford, J. M. 2012. Invited review: the impact of automatic milking systems on dairy cow management, behavior, health, and welfare. *Journal of Dairy Science* 95(5), 2227–47. doi:10.3168/jds.2011-4943.

Kawai, K., Hagiwara, S., Anri, A. and Nagahata, H. 1999. Lactoferrin concentration in milk of bovine clinical mastitis. *Veterinary Research Communications* 23(7), 391–8. doi:10.1023/A:1006347423426.

Kempthorne, O. and Nordskog, A. W. 1959. Restricted selection indices. *Biometrics* 15(1), 10–9. doi:10.2307/2527598.

Klungland, H., Sabry, A., Heringstad, B., Olsen, H. G., Gomez-Raya, L., Våge, D. I., Olsaker, I., Ødegård, J., Klemetsdal, G., Schulman, N., et al. 2001. Quantitative trait loci affecting clinical mastitis and somatic cell count in dairy cattle. *Mammalian Genome: Official Journal of the International Mammalian Genome Society* 12(11), 837–42. doi:10.1007/s00335001-2081-3.

Koeck, A., Heringstad, B., Egger-Danner, C., Fuerst, C., Winter, P. and Fuerst-Waltl, B. 2010. Genetic analysis of clinical mastitis and somatic cell count traits in Austrian Fleckvieh cows. *Journal of Dairy Science* 93(12), 5987–95. doi:10.3168/jds.2010-3451.

Koivula, M., Nousiainen, J. I., Nousiainen, J. and Mäntysaari, E. A. 2007. Use of herd solutions from a random regression test-day model for diagnostic dairy herd management. *Journal of Dairy Science* 90(5), 2563–8. doi:10.3168/jds.2006-517.

Kuhn, Ch, Bennewitz, J., Reinsch, N., Xu, N., Thomsen, H., Looft, C., Brockmann, G. A., Schwerin, M., Weimann, C., Hiendleder, S., et al. 2003. Quantitative trait loci mapping of functional traits in the German Holstein cattle population. *Journal of Dairy Science* 86(1), 360–8. doi:10.3168/jds.S0022-0302(03)73614-5.

Ledford, H. 2015. Salmon approval heralds rethink of transgenic animals. *Nature* 527(7579), 417–8. doi:10.1038/527417a.

Leitner, G., Shoshani, E., Krifucks, O., Chaffer, M. and Saran, A. 2000. Milk leucocyte population patterns in bovine udder infection of different aetiology. *Journal of Veterinary Medicine. B, Infectious Diseases and Veterinary Public Health* 47(8), 581–9. doi:10.1046/j.1439-0450.2000.00388.x.

Mark, T., Fikse, W. F., Emanuelson, U. and Philipsson, J. 2002. International genetic evaluations of Holstein sires for milk somatic cell and clinical mastitis. *Journal of Dairy Science* 85(9), 2384–92. doi:10.3168/jds.S0022-0302(02)74319-1.

Maxmen, A. 2017. Gene-edited animals face US regulatory crackdown. *Nature*. doi:10.1038/nature.2017.21331.

Nash, D. L., Rogers, G. W., Cooper, J. B., Hargrove, G. L., Keown, J. F. and Hansen, L. B. 2000. Heritability of clinical mastitis incidence and relationships with sire transmitting abilities for somatic cell score, udder type traits, productive life, and protein yield. *Journal of Dairy Science* 83(10), 2350–60. doi:10.3168/jds.S0022-0302(00)75123-X.

Nash, D. L., Rogers, G. W., Cooper, J. B., Hargrove, G. L. and Keown, J. F. 2002. Relationships among severity and duration of clinical mastitis and sire transmitting abilities for somatic cell score, udder type traits, productive life, and protein yield. *Journal of Dairy Science* 85(5), 1273–84. doi:10.3168/jds.S0022-0302(02)74192-1.

Nickerson, S. C. 1989. Immunological aspects of mammary Involution1. *Journal of Dairy Science* 72(6), 1665–78. doi:10.3168/jds.S0022-0302(89)79278-X.

Norberg, E. 2005. Electrical conductivity of milk as a phenotypic and genetic indicator of bovine mastitis: a review. *Livestock Production Science* 96(2–3), 129–39. doi:10.1016/j.livprodsci.2004.12.014.

Norberg, E., Hogeveen, H., Korsgaard, I. R., Friggens, N. C., Sloth, K. H. and Løvendahl, P. 2004. Electrical conductivity of milk: ability to predict mastitis status. *Journal of Dairy Science* 87(4), 1099–107. doi:10.3168/jds.S0022-0302(04)73256-7.

Nyman, A. -K., Emanuelson, U. and Waller, K. P. 2016. Diagnostic test performance of somatic cell count, lactate dehydrogenase, and N -acetyl-β- d -glucosaminidase for detecting dairy cows with intramammary infection. *Journal of Dairy Science* 99(2), 1440–8. doi:10.3168/jds.2015-9808.

Piepers, S., De Vliegher, S., Demeyere, K., Lambrecht, B. N., de Kruif, A., Meyer, E. and Opsomer, G. 2009. Flow cytometric identification of bovine milk neutrophils and simultaneous quantification of their viability. *Journal of Dairy Science* 92(2), 626–31. doi:10.3168/jds.2008-1393.

Pilla, R., Malvisi, M., Snel, G. G., Schwarz, D., König, S., Czerny, C. P. and Piccinini, R. 2013. Differential cell count as an alternative method to diagnose dairy cow mastitis. *Journal of Dairy Science* 96(3), 1653–60. doi:10.3168/jds.2012-6298.

Pryce, J. E. and Daetwyler, H. D. 2012. Designing dairy cattle breeding schemes under genomic selection: a review of international research. *Animal Production Science* 52(3), 107–14. doi:10.1071/AN11098.

Pryce, J. E., Veerkamp, R. F., Thompson, R., Hill, W. G. and Simm, G. 1997. Genetic aspects of common health disorders and measures of fertility in Holstein Friesian dairy cattle. *Animal Science* 65(3), 353–60. doi:10.1017/S1357729800008559.

Pyörälä, S. 2003. Indicators of inflammation in the diagnosis of mastitis. *Veterinary Research* 34(5), 565–78. doi:10.1051/vetres:2003026.

Rupp, R. and Boichard, D. 1999. Genetic parameters for clinical mastitis, somatic cell score, production, udder type traits, and milking ease in first lactation Holsteins. *Journal of Dairy Science* 82(10), 2198–204. doi:10.3168/jds.S0022-0302(99)75465-2.

Sahana, G., Guldbrandtsen, B., Thomsen, B. and Lund, M. S. 2013. Confirmation and fine-mapping of clinical mastitis and somatic cell score QTL in Nordic Holstein cattle. *Animal Genetics* 44(6), 620–6. doi:10.1111/age.12053.

Schepers, A. J., Lam, T. J. G. M., Schukken, Y. H., Wilmink, J. B. M. and Hanekamp, W. J. A. 1997. Estimation of variance components for somatic cell counts to determine thresholds for uninfected quarters. *Journal of Dairy Science* 80(8), 1833–40. doi:10.3168/jds.S0022-0302(97)76118-6.

Schöpke, K. and Swalve, H. H. 2016. Review: opportunities and challenges for small populations of dairy cattle in the era of genomics. *Animal: an International Journal of Animal Bioscience* 10(6), 1050–60. doi:10.1017/S1751731116000410.

Schrooten, C., Bovenhuis, H., Coppieters, W. and Van Arendonk, J. A. 2000. Whole genome scan to detect quantitative trait loci for conformation and functional traits in dairy cattle. *Journal of Dairy Science* 83(4), 795–806. doi:10.3168/jds.S0022-0302(00)74942-3.

Schukken, Y. H., Lam, T. J. G. M. and Barkema, H. W. 1997. Biological basis for selection on udder health. *Interbull Bulletin* 15, 27–33.

Schukken, Y. H., Wilson, D. J., Welcome, F., Garrison-Tikofsky, L. and Gonzalez, R. N. 2003. Monitoring udder health and milk quality using somatic cell counts. *Veterinary Research* 34(5), 579–96. doi:10.1051/vetres:2003028.

Schutz, M. M. 1994. Genetic evaluation of somatic cell scores for United States dairy cattle. *Journal of Dairy Science* 77(7), 2113–29. doi:10.3168/jds.S0022-0302(94)77154-X.

Schwarz, D. 2017. The new CombiFoss 7 DC. Differential somatic cell count and other advancements in milk testing. ICAR Tech. Series no. 22. International Committee for Animal Recording, Rome, Italy, pp. 41–7.

Schwarz, D., Diesterbeck, U. S., König, S., Brügemann, K., Schlez, K., Zschöck, M., Wolter, W. and Czerny, C. P. 2011. Flow cytometric differential cell counts in milk for the evaluation of inflammatory reactions in clinically healthy and subclinically infected bovine mammary glands. *Journal of Dairy Science* 94(10), 5033–44. doi:10.3168/jds.2011-4348.

Seegers, H., Fourichon, C. and Beaudeau, F. 2003. Production effects related to mastitis and mastitis economics in dairy cattle herds. *Veterinary Research* 34(5), 475–91. doi:10.1051/vetres:2003027.

Sewalem, A., Miglior, F. and Kistemaker, G. J. 2011. Short communication: genetic parameters of milking temperament and milking speed in Canadian Holsteins. *Journal of Dairy Science* 94(1), 512–6. doi:10.3168/jds.2010-3479.

Seykora, A. J. and McDaniel, B. T. 1985. Heritabilities of test traits and their relationships with milk yield, somatic cell count, and percent two-minute milk. *Journal of Dairy Science* 68(10), 2670–83. doi:10.3168/jds.S0022-0302(85)81152-8.

Shimazaki, K. I. and Kawai, K. 2017. Advances in lactoferrin research concerning bovine mastitis. *Biochemistry and Cell Biology = Biochimie et Biologie Cellulaire* 95(1), 69–75. doi:10.1139/bcb-2016-0044.

Sørensen, L. P., Mark, T., Madsen, P. and Lund, M. S. 2009. Genetic correlations between pathogen-specific mastitis and somatic cell count in Danish Holsteins. *Journal of Dairy Science* 92(7), 3457–71. doi:10.3168/jds.2008-1870.

Soyeurt, H., Bastin, C., Colinet, F. G., Arnould, V. M., Berry, D. P., Wall, E., Dehareng, F., Nguyen, H. N., Dardenne, P., Schefers, J., et al. 2012. Mid-infrared prediction of lactoferrin content in bovine milk: potential indicator of mastitis. *Animal: an International Journal of Animal Bioscience* 6(11), 1830–8. doi:10.1017/S1751731112000791.

Suriyasathaporn, W., Schukken, Y. H., Nielen, M. and Brand, A. 2000. Low somatic cell count: a risk factor for subsequent clinical mastitis in a dairy herd. *Journal of Dairy Science* 83(6), 1248–55. doi:10.3168/jds.S0022-0302(00)74991-5.

Thompson, R. 1980. A note on the estimation of economic values for selection indices. *Animal Science* 31(1), 115–7. doi:10.1017/S0003356100039842.

Thompson-Crispi, K. A., Sewalem, A., Miglior, F. and Mallard, B. A. 2012. Genetic parameters of adaptive immune response traits in Canadian Holsteins. *Journal of Dairy Science* 95(1), 401–9. doi:10.3168/jds.2011-4452.

Tiezzi, F., Parker-Gaddis, K. L., Cole, J. B., Clay, J. S. and Maltecca, C. 2015. A Genome-Wide Association Study for clinical mastitis in first parity US Holstein cows using single-step approach and genomic matrix Re-weighting procedure. *PLoS ONE* 10(2), e0114919. doi:10.1371/journal.pone.0114919.

VanRaden, P. M. 2017. Genomic tools to improve progress and preserve variation for future generations. Euro. Assoc. Anim. Sci. Annu. Mtg., Sess. 01, Managing Genet. Diversity in Cattle in the Era of Genomic Selec., Book of Abstr., p. 79 (abstr. S(01)_01).

VanRaden, P. M., Van Tassell, C. P., Wiggans, G. R., Sonstegard, T. S., Schnabel, R. D., Taylor, J. F. and Schenkel, F. S. 2009. Invited review: reliability of genomic predictions for North American Holstein bulls. *Journal of Dairy Science* 92(1), 16–24. doi:10.3168/jds.2008-1514.

VanRaden, P. M., Cole, J. B. and Parker Gaddis, K. L. 2018. Net merit as a measure of lifetime profit: 2018 revision. AIP Research Report NM$7, pp. 5–18. Available at: https://aipl.arsusda.gov/reference/nmcalc-2018.htm.

Vukasinovic, N., Bacciu, N., Przybyla, C. A., Boddhireddy, P. and DeNise, S. K. 2017. Development of genetic and genomic evaluation for wellness traits in US Holstein cows. *Journal of Dairy Science* 100(1), 428–38. doi:10.3168/jds.2016-11520.

Wall, R. J., Powell, A. M., Paape, M. J., Kerr, D. E., Bannerman, D. D., Pursel, V. G., Wells, K. D., Talbot, N. and Hawk, H. W. 2005. Genetically enhanced cows resist intramammary *Staphylococcus aureus* infection. *Nature Biotechnology* 23(4), 445–51. doi:10.1038/nbt1078.

Waltz, E. 2016. Gene-edited CRISPR mushroom escapes US regulation. *Nature* 532(7599), 293. doi:10.1038/nature.2016.19754.

Wellnitz, O. and Bruckmaier, R. M. 2012. The innate immune response of the bovine mammary gland to bacterial infection. *Veterinary Journal* 192(2), 148–52. doi:10.1016/j.tvjl.2011.09.013.

Wiggans, G. R., Thornton, L. L., Neitzel, R. R. and Gengler, N. 2007. Short communication: genetic evaluation of milking speed for Brown Swiss dairy cattle in the United States. *Journal of Dairy Science* 90(2), 1021–3. doi:10.3168/jds.S0022-0302(07)71587-4.

Wiggans, G. R., Cole, J. B., Hubbard, S. M. and Sonstegard, T. S. 2017. Genomic selection in dairy cattle: the USDA experience. *Annual Review of Animal Biosciences* 5(1), 309–27. doi:10.1146/annurev-animal-021815-111422.

Zhang, W. C., Dekkers, J. C., Banos, G. and Burnside, E. B. 1994. Adjustment factors and genetic evaluation for somatic cell score and relationships with other traits of Canadian Holsteins. *Journal of Dairy Science* 77(2), 659–65. doi:10.3168/jds.S0022-0302(94)76996-4.

Zottl, K. 2016. On-farm recording of novel traits – genetic parameters and recommendations. ICAR Technical Series no. 20. In: *40th ICAR Biennial Session*, Puerto Varas, Chile. International Committee for Animal Recording, Rome, Italy.

Zwald, N. R., Weigel, K. A., Chang, Y. M., Welper, R. D. and Clay, J. S. 2004. Genetic selection for health traits using producer-recorded data. I. Incidence rates, heritability estimates, and sire breeding values. *Journal of Dairy Science* 87(12), 4287–94. doi:10.3168/jds.S0022-0302(04)73573-0.

Chapter 4

Minimizing the development of antimicrobial resistance on dairy farms: appropriate use of antibiotics for the treatment of mastitis[1]

Pamela L. Ruegg, University of Wisconsin-Madison, USA

1 Introduction

Antibiotics are an essential tool for combatting bacterial diseases and their use has contributed to increased welfare of both human beings and animals. The discovery of penicillin in the late 1920s revolutionized medicine and the first veterinary use of this compound is reported to have been for treatment of mastitis in dairy cows (Mitchell et al., 1998). Antibiotics came into common use beginning in the late 1940s and concerns about emergence of antimicrobial resistance were noted almost immediately (Prescott, 2006). Initially, concerns about the development of resistance were minimal because new classes of drugs were continually being developed, but contemporary researchers have noted the gradual emergence of resistance to all classes of drugs and resistance mechanisms have been identified for all antibiotics (Boerlin and White, 2013). The issue of antibiotic

1. The terms 'antimicrobial' and 'antibiotic' are used interchangeably in this paper but are not synonymous. In technical terms, antibiotics refer only to substances of microbial origin (such as penicillin) that are active against other microbes, while antimicrobial refers to any substance (including synthetic compounds) that destroys microbes.

http://dx.doi.org/10.19103/AS.2016.0005.22

resistance has gained attention globally and this issue will continue to be relevant for dairy farmers and the veterinarians who advise them.

Use of antibiotics in animal agriculture is under increasing scrutiny, especially in wealthy countries where few citizens are familiar with farm management practices. In response to consumer concerns, antibiotic choices may be limited by supplier codes imposed by multinational food companies or by governmental regulations. In most countries, concerns about use of antibiotics have been partially addressed by limiting agricultural uses to antimicrobials that are less critical for human health needs; however, classes approved for use in animals vary among countries and species. While there is no centralized international authority that tracks agricultural uses of antimicrobials, European researchers have documented considerable differences among countries in the quantity of antimicrobials used for production of 1 kg of meat (Garcia-Migura et al., 2014). This type of comparative data is not available for dairy farms, but compared with other food animal species, restrictions on sale of milk from cows receiving antibiotics have resulted in less non-therapeutic uses of antibiotics on dairy farms. However, therapeutic use of antimicrobials in dairy cows has the potential to affect human health by increasing the risk of exposure to antimicrobial residues in foodstuffs (Ruegg and Tabone, 2000) or by influencing generation or selection of resistant pathogens. In countries where the dairy sector has developed, the risk of exposure to antimicrobial residues has been addressed through the use of effective regulatory mechanisms, but there is increasing concern about the role of antimicrobial usage in the development of antimicrobial resistance of microorganisms that may contaminate food or the environment (Sandberg and LaPara, 2016). For many dairy products, these risks are somewhat mitigated by post-harvest practices (such as pasteurization) that reduce the probability of exposure to farm-related pathogens. In fact, in less developed regions, post-harvest contamination of dairy products with resistant pathogens of human origin may pose a greater risk than exposure to resistant organisms that could originate from animals (Al-Ashmawy et al., 2016; Schmidt et al., 2015).

Worldwide, mastitis is the most prevalent bacterial disease of dairy cows and the use of antimicrobials for the control of this disease is of concern because most antimicrobials given to dairy cows are for treatment or prevention of this disease. In response to concerns about antimicrobial usage on farms, some countries have enacted legislation that requires reduced usage of antimicrobials (Kuipers et al., 2016). These types of legislation are typically based on assumptions that antimicrobials are used excessively and that reducing antimicrobial usage will result in decreased proportions of resistant organisms and reduced threats to human health. While there is no compelling evidence that the use of antimicrobials for treatment of mastitis has resulted in increased prevalence of resistant pathogens (Oliver and Murinda, 2012; Erskine et al., 2004), ensuring continued efficacy of antimicrobials is a public health priority and judicious usage of antimicrobials in animal agriculture is a societal obligation that must be met by the dairy industry. The objective of this chapter is to review how antibiotics are used on dairy farms and offer recommendations that will help to minimize the unnecessary use of antimicrobials for treatment of mastitis.

2 Use of antimicrobials on dairy farms

Throughout the world, antimicrobials are primarily used to treat diseases that have a bacterial aetiology, and on dairy farms, most antimicrobials are used therapeutically but some are used

Table 1 Comparison of estimated antimicrobial usage in studies standardizing usage on dairy farms using defined daily doses (DDD)

Year	Pol and Ruegg, 2007b	Saini et al., 2012a	Gonzalez-Pereyra et al., 2015	Kuipers et al., 2016	Stevens et al., 2016
Country	WI, USA	Canada	Argentina	Holland	Belgium
Number of herds	40[a]	89	18	94	57
Number of lactating cows/herd	197	69	219	94	69
Total defined daily doses per cow per year	5.43	5.24[b]	5.21	5.45[c]	7.6[b]
% intramammary route (of total)	66	35	85	72[d]	63

[a]Antibiotic usage reported only for enrolled 20 conventional herds.
[b]These studies reported DDD per 1000 cow-days, which was converted to per cow/yr by dividing by 2.74 (1000 cow-days/365).
[c]Average over 7-yr period of surveillance.
[d]In 2012 (last year of surveillance).

to prevent disease during periods of increased susceptibility. For example, in the United States, treatments for mastitis, lameness, reproductive disorders, digestive disorders and respiratory disease are the most common reasons that adult dairy cows receive antimicrobials (USDA, 2008) and it is likely that a similar situation exists in other developed dairy regions. While methodological differences make it difficult to compare antimicrobial usage among studies conducted in different regions, quantification of usage can be standardized by the calculation of defined daily dosages (DDDs) (Jensen et al., 2004). Using this method, several studies have confirmed that treatment of mastitis accounts for the majority of antibiotics used on dairy farms (Table 1) (USDA, 2008; Pol and Ruegg, 2007b; Saini et al., 2012a; Gonzalez Pereyra et al., 2015; Stevens et al., 2016; Kuipers et al., 2016). While the calculation of DDD does vary among studies (based on dosages, animal size and definition of dry cow therapy), most studies have reported that dairy cows receive about 5–8 DDD of antibiotics per cow per year and about 35–85% of the doses are administered via intramammary (IMM) infusion. Depending on drug approvals and regulations, in some countries, parenteral treatments are also frequently given for treatment of mastitis. Even in the United States, where no antibiotics are approved for parenteral treatment of mastitis (but extralabel usage of some drugs is allowed when prescribed by a veterinarian), parenterally administered antimicrobials used for treatment of mastitis accounted for about half of all parenteral usage of antimicrobials and 17% of total usage (Pol and Ruegg, 2007b).

Mastitis is the most common reason for antibiotic usage because it is the most prevalent bacterial disease of adult dairy cows, and most farmers believe that antibiotics should be used to treat this disease (Jones et al., 2015). In a recent study of large dairy farms in Wisconsin (Oliveira and Ruegg, 2014), all farms (n = 51) reported antibiotic treatments for clinical mastitis and reproductive disorders, while fewer farms reported the use of antibiotics for treatment of respiratory disease (90% of farms), lameness (82%) or digestive disorders (32%). However, a much greater incidence of mastitis results in a dramatically greater proportion of antibiotic usage being attributed to this disease. For example, there were 40 treatments/100 cows/yr for mastitis as compared to 13 treatments/100 cows/yr for reproductive disorders and <5 treatments per 100 cows/yr for respiratory disease,

Table 2 Frequency of usage (ranking) of top five antimicrobials used in adult cows on dairy farms for selected countries

	Italy	Holland	Canada	USA	Belgium
	Serraino et al., 2013	Wageningen, 2012[a]	Saini et al., 2012a	NAHMS, 2008	Stevens et al., 2016
Combinations	1[b]	2[c]	3[d]		4[e]
Ampicillin or cloxacillin	2			5	
Other antimicrobials	3	5			
Enrofloxacin	4				
Cephalosporin – 1st					5
Cephalosporin – 3rd	5		2	1	3
Cephalosporin – 4th					1
Oxytetracycline or tetracycline		3	4	2	
Lincomycin and spectinomycin					
Sulphonamides				4	
Penicillin		1	1	3	2
Trimethoprim and sulphonamides		4	5		

[a]http://www.wageningenur.nl/en/Research-Results/Projects-and-programmes/MARAN-Antibiotic-usage/Introduction.htm.
[b]Amoxicillin/clavulanic acid, rifaximin/cephacetril, cloxacillin/ampicillin.
[c]Amoxicillin/colistin, cephalexin/kanamycin, penicillin/streptomycin, lincomycin/neomycin, penicillin/neomycin.
[d]Penicillin combinations.
[e]First generation cephalosporin/aminoglycoside.

lameness or digestive problems (Ruegg, data not shown in published study). This data indicates that efforts to reduce antimicrobial usage on the dairy farm must be targeted on the prevention and appropriate treatment of mastitis.

Throughout the world, drug usage on dairy farms is increasingly restricted, but regulations governing allowable drug usage are not consistent and mastitis treatments vary enormously among countries (Table 2). In some countries (such as Italy and Argentina), IMM products containing combinations of antibiotics are frequently administered (Serraino et al., 2013; Gonzalez Pereyra et al., 2015), while in other countries, most combination products are not approved (Oliveira and Ruegg, 2014). In the United States, no antimicrobials are approved for parenteral treatment of mastitis, only two antimicrobial classes are represented among commercially available IMM products, usage of fluoroquinolones and fourth-generation cephalosporins is prohibited and sulphonamide use is highly restricted. In contrast, fluoroquinolones and fourth-generation cephalosporins are approved and commonly used in some countries. While a variety of drugs are used throughout the world, data consistently demonstrate that β-lactam compounds are the most common antimicrobial class administered to dairy cows (Table 2). It is interesting to note that, while different antibiotics are used for treatment of mastitis in different countries, there is no scientific evidence that therapeutic outcomes vary among countries. As the overwhelming usage of antimicrobials is for the treatment of mastitis, efforts to reduce antimicrobial usage must be focused on appropriate treatment of this disease. Evidence to support the efficacy of many commonly used treatment protocols is mostly anecdotal and scientifically validated therapeutic protocols for the treatment of mastitis and other infectious dairy diseases are sorely needed.

3 Clinical relevance of antimicrobial resistance data

Bacterial resistance to specific antimicrobial classes may occur intrinsically (due to lack of binding sites or other pharmacological characteristics) and can cause treatment failure but is not considered a major public health issue (Neu, 1992). Intrinsic resistance usually occurs at the genus or species level and one example is a Gram-negative bacterium that has an outer membrane that is impermeable to the chosen antibiotic (such as pirlimycin). Acquired resistance occurs when a previously susceptible bacterium becomes resistant through mutation or acquisition of new DNA. Acquired resistance is strain specific and the presence of the antibiotic will subsequently select for resistant strains (Prescott, 2006). Acquired resistance has the potential for transmission to humans and is of great concern to public health authorities (Neu, 1992). The ability of organisms to acquire resistance varies among antimicrobials. For example, researchers routinely report greater proportions of Gram-positive mastitis organisms that are resistant to pirlimycin as compared to resistance to first-and third-generation cephalosporins (Ruegg et al., 2015a). This resistance is generally acquired and also affects other antibiotics within the macrolide, lincosamide and streptogramin grouping. Knowledge of the ability and likelihood of a particular antimicrobial class to induce resistance should be considered when veterinarians recommend mastitis therapies.

Resistance to antimicrobials is typically determined using phenotypic tests that measure the ability of an antimicrobial to inhibit bacterial growth through *in vitro* diffusion or dilution tests. The determination of phenotypic resistance is based on achieving an inhibitory concentration of the drug that is greater than a specified breakpoint (antimicrobial concentration). Few antimicrobial breakpoints for mastitis pathogens are clinically validated and associations between results of phenotypic sensitivity tests and clinical outcomes are weak and vary among organisms and drugs (Apparao et al., 2009a,b; Hoe and Ruegg, 2005), emphasizing the importance of host factors in the elimination of IMM infection (IMI). Surveillance studies have primarily examined the occurrence of phenotypic resistance, but researchers are increasingly focusing on the identification of resistance genes. The presence of resistance genes in an organism does not always correspond to the expression of resistance phenotypes. For example, an organism may contain blaZ (the gene that encodes β-lactam resistance) or another gene but be phenotypically sensitive (Ruegg et al., 2015a; Haveri et al., 2005). Thus, associations of resistance genes with expectations of positive or negative clinical outcomes should be made cautiously and more research is needed to validate the clinical usefulness of resistance breakpoints for veterinary drugs. In many regions, there is little variability in the results of susceptibility tests of mastitis pathogens and these results are often less useful for predicting clinical outcomes than knowledge of the aetiology and a review of the cow's medical history.

4 Trends in the antimicrobial resistance of mastitis pathogens

Exposure to antimicrobials is well known to select for resistant organisms, but the evolution and maintenance of resistant mastitis pathogens in dairy cows or dairy farm environments has not been well described. At least one researcher has reported that the level of antibiotic resistance genes in soils that received manure from cows that were treated therapeutically with antibiotics rapidly return to background levels (Sandberg and LaPara, 2016). A recent paper reviewed the prevalence of antimicrobial resistance of mastitis pathogens using studies from throughout the world and concluded that there is relatively little evidence to suggest that widespread resistance is emerging or progressing (Oliver and Murinda, 2012). Antimicrobial resistance of *Staphylococcus aureus* isolated from bovine mastitis cases has been extensively studied. While methicillin-resistant *S. aureus* is generally rare on specialized dairy farms (Gindonis et al., 2013; Tenhagen et al., 2014; Cicconi-Hogan et al., 2014), researchers have demonstrated dramatically different proportions of *S. aureus* that were resistant to antibiotics commonly used on dairy farms (Intorre et al., 2013; Moroni et al., 2006; Saini et al., 2012b; Oliveira et al., 2012; Thomas et al., 2015; Petrovski et al., 2015; Bengtsson et al., 2009) (Table 3). For example, in recent years, *S. aureus* isolated from bovine mastitis occurring in cows in North America have typically demonstrated little phenotypic resistance (Saini et al., 2012b; Oliveira et al., 2012; Ruegg et al., 2015b), while studies conducted in some European countries have reported more resistance (Thomas et al., 2015; Intorre et al., 2013; Moroni et al., 2006) and Intorre et al. (2013) reported a significant increasing trend in resistance to several important antimicrobials used to treat mastitis on Italian dairy farms.

From a research standpoint, it is difficult to determine if there is an association between mastitis treatments and trends in the resistance of mastitis pathogens because quantifying exposure to antimicrobials and associating that exposure with resistance is very challenging. Some studies have attempted to describe differences in the susceptibility of isolates obtained from farms with differing histories of exposure to selected antimicrobials (Pol and Ruegg, 2007a; Saini et al., 2012c, 2013). Using this approach, researchers have consistently demonstrated variation in resistance to some drugs among farms and among organisms and some associations of antimicrobial resistance with antimicrobial usage on farms. Interestingly, the greatest variation in resistance is based on the class of the drug, rather than exposure to drugs. Greater exposure to some commonly used antimicrobials has been linked to a greater proportion of resistant organisms, but increased exposure to other commonly used antimicrobials has not been associated with greater resistance. However, retrospective studies and reviews have reported little evidence of a systematic increase in resistance associated with drugs used for the treatment and prevention of mastitis (Erskine et al., 2002; Makovec and Ruegg, 2003; Oliver and Murinda, 2012). While continued research is needed, it is evident that limiting the potential for the development of resistance is a priority, and veterinarians and dairy producers must have justification for the use of antibiotics and provide evidence-based recommendations for the treatment of mastitis.

Table 3 Proportion of *Staphylococcus aureus* isolates resistant to selected antimicrobials

	Erskine et al., 2002[a]	Makovec and Ruegg, 2003[a]	Moroni, et al., 2006	Bengtssen et al., 2009	Saini et al., 2012b[a]	Oliveira et al., 2012	Intorre et al., 2013	Petrovski et al., 2015[a]	Thomas et al., 2015
Location	MI, USA	WI, USA	Italy	Sweden	Canada	WI, USA	Italy	New Zealand	8 European countries
Farms (*n*)	>100	>100	42	–	91	13	>100		–
Isolates (*n*)	852	2132	68	211	562	116	1200	107	250
Period	1994–2000	1994–2001	2004	2002–3	2007–8	2010	2005–11	2006–7	2002–6
Ampicillin	50%	35%	99%	–	3%	5%	–	20%	–
Ceftiofur	<1%	–	–	–	<1%	0%	–	–	0%
Cefquinome	–	–	–	–	–	–	5%	–	–
Cephalothin	<1%	<1%	–	0%	0%	2%	8%	0%	0%
Cloxacillin	–	2%	0%	–	0%	–	–	0%	0%
Erythromycin	7%	7%	–	2%	<1%	5%	43%	0%	<1%
Penicillin	50%	35%	69%	7%	9%	7%	65%	20%	36%
Pirlimycin	2%	5%	–	–	2%	4%	–	–	–
Tetracycline	9%	9%	59%	0%	3%	10%	25%	0%	5%
Pen/Novo	–	<1%	–	–	<1%	2%	–	–	–
Oxacillin	<1%	–	0%	0%	0%	0%	13%	0%	0%
Trimethoprim/ sulpha	<1%	<1%	–	0%	–	–	3%	–	–

[a]Intermediate and resistant are combined.

5 Ensuring effective use of antibiotics in the treatment of mastitis: diagnosis, antibiotic choice and duration of treatment

5.1 Consistent detection, diagnostic protocols and recording

Principle 1: Dairy producers should work with veterinarians to ensure consistent detection and diagnostic protocols as well as the development of recording systems that facilitate the ability to evaluate the results of mastitis treatments.

Clinical mastitis is technically defined as the production of abnormal milk with or without secondary symptoms, but the working definition of clinical mastitis varies greatly among farm personnel. On large farms, detection of mastitis is usually dependent on the training and observational skills of the milking technicians. Veterinarians must actively communicate with milking technicians and farm managers to be sure that the definition of clinical mastitis and the intensity of detection are consistent with farm goals. Case definitions for mastitis should be simple and easily understood by all farm workers. Mastitis severity scores should be recorded for each case in permanent cow treatment records. The use of a three-point severity scoring scale (1 = abnormal milk only; 2 = abnormal milk and abnormal udder; 3 = abnormal milk accompanied by symptoms that extend beyond the udder) is practical, is simply recorded and can be an important method to monitor clinical mastitis detection intensity (Pinzon-Sanchez and Ruegg, 2011). In general, about 50% of clinical cases present with the mildest severity and cannot be detected unless foremilk is examined (Oliveira et al., 2013). When mild and moderate cases are not detected, the proportion of severe cases is often overestimated. When using this scale, if the proportion of severe cases exceeds about 5–15% of all cases, it is a signal that detection intensity and case definition should be investigated. Classification of mastitis based on severity allows for immediate treatment of severe cases while allowing time for review of the medical history of cow and determination of aetiology prior to treatment of non-severe cases.

It is difficult for farm personnel and veterinarians to determine if mastitis treatments are effective as bacteriological cure is not generally evaluated and clinical outcomes are often misleading. The appearance of milk is the most obvious symptom of clinical mastitis and is a result of the inflammatory process associated with the immune response to IMI. Visible signs of inflammation are not necessarily equivalent to the presence of an active IMI and generally resolve (with or without bacteriological cure) in approximately 4–6 days; so the appearance of milk should not be used to judge the efficacy of treatments (Oliveira and Ruegg, 2014). Longer-term outcomes should be evaluated and the rate of recurrence of clinical mastitis (<10% same quarter recurrence within 60–90 days) and somatic cell count (SCC) reduction (return to SCC <250 000 cells/mL by 60 days) should be routinely evaluated (Ruegg, 2011; Pinzon-Sanchez and Ruegg, 2011).

5.2 Choice of antibiotic

Principle 2: The choice of the antibiotic should be appropriate for the aetiology and narrow-spectrum drugs are preferred as the first choice.

Antibiotics are classified by the World Health Organization based on their importance for treating human illnesses (Anonymous, 2012). In developed dairy regions, most approved IMM antibiotics are not classified as high-priority drugs for treatment of human illnesses, but third-and fourth-generation cephalosporins, fluoroquinolones and macrolides are listed as high priority and critically important for human health. In regions where some of these drugs are approved for use for treatment of mastitis, these drugs should be reserved for cases where the efficacy of narrower-spectrum drugs is not expected. Based on intrinsic susceptibility of the target pathogens, antibiotics are classified as narrow or broad spectrum. Narrow-spectrum antibiotics have activity against either Gram-positive or Gram-negative bacteria, while broad-spectrum antibiotics have activity against both types of organisms. Narrow-spectrum drugs are usually less critical for human health needs (Weese et al., 2013) and should be the first choice for treatment of mastitis as they have less potential for selection for resistance. Generally, all over the world there are many approved IMM products that are considered as narrow-spectrum drugs, and therefore, the use of broader-spectrum IMM drugs should be reserved for cases where the aetiology is known and where narrow-spectrum drugs are not expected to be efficacious.

Drug regulations vary among countries. In some countries, only veterinarians are allowed to administer antibiotic treatments and all drugs used must be approved for that use. In other countries, farmers may administer some drugs that are explicitly approved for a particular usage and some non-approved usages are allowed under veterinary supervision. This 'extralabel use' includes administration for durations or dosing intervals that are not explicitly listed on the product label. Regardless of the country, veterinarians should be actively involved in developing and evaluating mastitis treatment protocols, and when deviations from approved antibiotic labels are proposed, these treatments should be performed under veterinary supervision.

5.3 Duration of use

Principle 3: Unless the aetiology or medical history of the cow suggests that longer duration antibiotic therapy will be beneficial, antibiotics should be used for as short a duration as possible.

Abnormal appearance of milk is a non-specific sign of inflammation that is not always predictive of ongoing IMI or aetiology. Neither the antibiotic choice nor the duration of treatment should be based solely on the appearance of milk. The appropriate duration of antibiotic treatment for CM is not well defined and it varies according to the aetiology. There is considerable evidence that longer duration antibiotic therapy increases bacteriological cure of pathogens that have the ability to penetrate into mammary tissue (such as *S. aureus* and some environmental *Streptococci* spp.) (Oliver et al., 2004a,b). However, research has not demonstrated that longer duration therapy results in improved clinical outcomes of infections caused by pathogens that tend to localize on superficial mucosal surfaces (such as coagulase-negative *Staphylococcus* spp. or most *Escherichia coli*). Modelling of the economic impact of non-specific treatments suggests that longer duration therapy should not be indiscriminately used as it significantly increases costs without improving economic outcomes (Pinzon-Sanchez et al., 2011). Mastitis is caused by a diverse group of bacteria and antibiotic therapy is not indicated for a large proportion of aetiologies. When indicated based on aetiology, the duration of therapy should be extended, but when the cause is not known, short-duration treatment is recommended. It is also important to assess the ability of farm workers to perform aseptic IMM infusions as extended IMM treatment is associated with an increased risk of infection from opportunistic pathogens, and herds with poor infusion techniques are not good candidates for multiple doses of IMM tubes.

6 Ensuring effective use of antibiotics in the treatment of mastitis: targeting treatment

6.1 Cow medical history

Principle 4: Treatments should be administered only after the medical history of the cow has been evaluated to determine if antibiotics will be of any benefit.

In a number of countries, most mastitis treatments are administered by farmers based on observation of abnormal milk and many of these treatments may be given to cows that are not likely to respond to therapy. Antibiotics are administered to aid the cow's immune system in the elimination of IMI and not all cows are good candidates for treatment. A number of characteristics of the cow are known to influence the probability of successful immune response (Burvenich et al., 2003) and many of those same characteristics are associated with the likelihood that an IMI will be eliminated by treatment. As cows age, the risk of both subclinical and clinical mastitis increases and numerous studies have indicated that older cows have poorer responses to antibiotic therapies as compared to younger cows (Sol et al., 2000; Hektoen et al., 2004; Barkema et al., 2006). It is well known that the bacteriological cure of older cows that received antibiotics for the treatment of mastitis is less than that of younger cows (McDougall et al., 2007) and age has also been associated with reduced clinical responses to therapy. Hektoen et al. (2004) measured responses to treatment by comparing scores for both acute and chronic symptoms obtained before treatment and at various periods after treatment. While parity was not associated with differences in acute symptoms, the reduction in chronic symptoms (changes in the milk, gland or inflammatory response) was markedly greater in first lactation than in older cattle. On a practical basis, older cows are more likely to be chronically infected with persistent subclinical infections that may result in periodic clinical episodes. These chronic infections are often refractory to antibiotic therapy and repeated use of antibiotics for clinical cases occurring in these quarters cannot be justified. The duration of subclinical infection prior to a clinical case is also associated with prognosis (Bradley and Green, 2009; Pinzon-Sanchez and Ruegg, 2011) and cows with a long history of high SCC have a reduced probability of achieving successful treatment outcomes. Likewise, cows with a history of repeated treatments for clinical mastitis are not likely to benefit from additional therapies, and therefore, the administration of antibiotics in such cows contributes to unjustifiable antibiotic usage. Before the administration of antibiotics, cow-level factors should be evaluated using individual cow health records that include monthly SCC values. For cows that may not benefit from antibiotics, 'watchful waiting' should be considered as the immediate case management option. Watchful waiting consists of isolation of the cow and discard of the abnormal milk until it returns to normal. In most instances, visible inflammation will subside within about 4–6 days. It is important for farm managers to recognize that the return to normal appearance of milk does not equate with 'cure' of IMI, and therefore, longer-term options, such as segregation of the cow, culling or cessation of lactation in an individual quarter (if a single quarter is repeatedly affected), should be considered.

6.2 Ensuring antibiotics are used only against bacterial infections

Principle 5: Antibiotics should be used only when there is a reasonable likelihood that a bacterial infection that can be effectively treated with available antibiotics is present.

Widespread adoption of recommended best management practices has allowed many farms to successfully control *S. aureus* and *Streptococcus agalactiae* and on modern dairy farms the distribution of pathogens causing clinical mastitis is quite diverse and often reflects environmental exposures (Table 4). Virulence, pathogenesis and prognosis of IMI are influenced by important characteristics that vary among pathogens. Depending on these characteristics, bacteria infect different areas of the gland, vary in their ability to cause systemic symptoms, have differing subclinical phases and differ in prognosis after treatment. In regions where confinement housing is used, *E. coli* is commonly cultured from clinical cases and about two-thirds of these cases present with symptoms that are localized to the udder (Table 4) (Oliveira et al., 2013). *E. coli* often infect superficial mucosal surfaces of the mammary gland and spontaneous bacterial clearance of this pathogen is common, so in many instances, antibiotic therapy is not necessary. In contrast, in regions that use pasture (such as New Zealand), mastitis is often caused by environmental *Streptococci* spp., of which some species (i.e. *Streptococcus uberis*) have the ability to deeply invade mammary gland secretory tissue (Table 4). Spontaneous cure of these organisms is not common and antimicrobial therapy is normally recommended.

Most researchers report that milk samples obtained from about 25–40% of clinical cases are microbiologically negative (before treatment). Many of these cases are apparent spontaneous cures before the symptoms are detected and the prognosis for these cases is excellent. In a study conducted in WI, about 75% of culture negative clinical cases were classified as bacteriologically cured after treatment (all received non-specific IMM antibiotic treatment), while in contrast, the odds of bacteriological cure for clinical cases caused by Gram-positive organisms were reduced by about 50–85% (Oliveira and Ruegg, 2014). Similarly, the cow-level recurrence rate of clinical mastitis within 60 days was 11% (n = 123) versus 20% for culture negative cases, and 35% for cows with cases caused by

Table 4 Results of selected studies that describe the distribution of bacteria recovered from milk of cows with clinical mastitis in modern dairy herds located in developed countries

Country	Herds	Milk samples[a]	*S. aureus*	Other staph	*S. agalactiae*	Other strep	Coliform	Other	No growth
Holland (de Haas, 2002)	274	2,737	18%	6%	0%	25%	28%	NR[b]	22%
UK (Bradley et al., 2007)	90	480	3%	13%	0%	25%	21%	11%	27%
New Zealand (McDougall et al., 2007)	28	1,332	19%	7%	0%	45%	NR	4%	27%
Canada (Olde Riekerink, 2007) (Riekerink et al., 2008)	106	2,850	11%	6%	0%	16%	15%	5%	47%
USA (Oliveira et al., 2013)	50	741	3%	7%	0%	11%	36%	16%	27%

[a]Results characterized as contaminated and mixed infections were excluded.
[b]NR indicates that the study did not report that outcome.
Source: Adapted from Ruegg et al., 2014.

Gram-positive (n = 128) or Gram-negative cases, respectively (data not shown in published study).

Knowledge of the aetiology is essential for making an informed decision about the usefulness of antibiotic therapy. Expectations for a spontaneous bacteriological cure of subclinical and clinical mastitis caused by *S. aureus* are essentially nil (Oliver et al., 2004b), while the expectation for a spontaneous cure of mastitis caused by *E. coli* is quite high (Suojala et al., 2013). For other pathogens, (yeasts, *Pseudomonas aeruginosa, Mycoplasma bovis, Prototheca zopfii, Serratia* spp. etc.), the use of antimicrobials for treatment cannot be recommended as the spectrum of approved drugs does not extend to these genera and there are virtually no clinical trials that can be used to justify appropriate antimicrobial usage. Mastitis is detected based on the observation of non-specific symptoms of inflammation and the organism cannot be determined without laboratory testing. Determination of the aetiology of IMI either before initiating therapy or to modify therapy is recommended and can considerably reduce unnecessary antimicrobial usage.

With current laboratory methods, it is not feasible for all farms to determine the aetiology of clinical cases before beginning therapy, and if the aetiology is not known, short-duration therapy with an approved IMM antibiotic is recommended (Pinzon-Sanchez et al., 2011). For larger herds (>200 cows), guiding treatment by the use of on-farm culture (OFC) systems has been shown to be economically beneficial (Lago et al., 2011a,b). On most farms, OFC methods are based on the use of laboratory shortcuts and have a goal of rapidly reaching a presumed diagnosis to guide treatment. Growth on a selective media is used to differentiate cases as caused by Gram-positive or Gram-negative bacteria, culture-negative cases or in some instances specific pathogens. After 24 hours of incubation, culture plates are observed and the treatment protocol is specified based on the culture outcome. Studies have indicated that 24-hour interpretation of selective agars used in OFC systems is about 80% accurate in differentiating Gram-positive and Gram-negative pathogens as compared to diagnostic laboratories (Lago et al., 2011a). Most smaller herds (<200 cows) do not have sufficient cases of mastitis to develop the expertise needed for OFC and an alternative is to offer rapid culturing using selective media at a local veterinary clinic. In these instances, farmers usually collect a milk sample and may immediately initiate treatment. After 24 hours of incubation, the veterinary clinic can send an email or text message with the preliminary microbiological diagnosis and instructions for modifications to the treatment. For example, if the culture result is microbiologically negative or Gram-negative, treatment may be stopped, while if the result is Gram-positive the veterinarian may recommend that the duration or drug be modified. Use of OFC to direct treatment of clinical mastitis gives farmers the opportunity to make better treatment decisions and will also reduce costs associated with milk discard and treatment of Gram-negative and microbiologically negative cases. Selective antibiotic treatment based on the use of OFC has not been demonstrated to reduce animal well-being. A positively controlled clinical trial evaluating OFC demonstrated that there were no significant differences in either long-term or short-term outcomes for cases of mastitis that received treatment based on the results of OFC as compared to cases treated immediately without regard to diagnosis (Lago et al., 2011a,b). In this study, antimicrobials were not administered to cases that were culture negative or Gram negative; thus, the use of IMM antimicrobials was reduced by approximately 50% as compared to cases, which were treated without prior diagnosis. The use of selective treatment protocols based on rapid culture methodologies can result in reductions in unnecessary antibiotic treatments, but it is important for dairy farmers and veterinarians to recognize that the methods used in OFC laboratories are not equivalent to those in diagnostic laboratories, and overseeing of the results is necessary to ensure that mistakes do not occur.

7 Conclusions

Mastitis is the most common bacterial disease of dairy cows and is the most common reason that antibiotics are administered to adult cows. While there is no overt evidence that the use of antibiotics for treatment of mastitis is resulting in emerging resistance to medically important antibiotics, there is much opportunity for improvement in mastitis treatments, and the justifiable usage of antibiotics is a societal obligation that the worldwide dairy industry must strive to achieve. Tremendous progress has been made in the prevention of mastitis and continued emphasis must be placed on reducing the incidence of this disease. However, when animals do become infected, veterinarians should be involved in developing and implementing mastitis treatment protocols that include justifiable usage of antibiotics. Research evidence is available to help guide mastitis treatment decisions and to better select animals that will benefit from specific treatments. There is sufficient research evidence to help develop mastitis treatment protocols that vary according to animal characteristics and the history of subclinical disease. Determination of aetiology is one of the most important steps in justifying antibiotic treatment. The use of rapid culturing programmes is recommended to guide selective treatment programmes. Appropriate and justifiable usage of antibiotics for treatment of mastitis is necessary to maintain animal well-being, but should be guided by principles that minimize the possibility of inappropriate usage.

8 Where to look for further information

Information about the prudent use of antimicrobial agents can be found at http://www.fil-idf.org/Public/Download.php?media=40125.

Comprehensive mastitis control information can be found at the website of the National Mastitis Council (www.nmconline.org).

Videos about pathogen-specific preventive strategies and the development and use of OFC programmes to guide selective treatment can be found at the UW Madison Milk Quality website (http://milkquality.wisc.edu).

Active mastitis research and outreach programmes can be found all over the world. Several prominent English language programmes include:

Canadian Bovine Mastitis Research Network: http://www.medvet.umontreal.ca/reseau_mammite/en/index.php.

Quality Milk Production Service (Cornell University): https://ahdc.vet.cornell.edu/sects/QMPS/.

Countdown Down Under (Dairy Australia): http://www.dairyaustralia.com.au/Animal-management/Mastitis.aspx.

DairyNZ (New Zealand) – resources for seasonal and pasture-based dairy farms: http://www.dairynz.co.nz/animal/mastitis/tools-and-resources/.

9 References

Al-Ashmawy, M. A., Sallam, K. I., Abd-Elghany, S. M., Elhadidy, M. and Tamura, T. 2016. Prevalence, molecular characterization, and antimicrobial susceptibility of Methicillin-Resistant Staphylococcus aureus (MRSA) isolated from milk and dairy products. *Foodborne Pathog Dis.* 13, 156–62.

Anonymous. 2012. Critically important antimicrobials for human health – 3rd rev. (Accessed 20 January 2016).

Apparao, D., Oliveira, L. and Ruegg, P. L. 2009a. Relationship between results of in vitro susceptibility tests and outcomes following treatment with pirlimycin hydrochloride in cows with subclinical mastitis associated with gram-positive pathogens. *Journal of the American Veterinary Medical Association*, 234, 1437–46.

Apparao, M. D., Ruegg, P. L., Lago, A., Godden, S., Bey, R. and Leslie, K. 2009b. Relationship between in vitro susceptibility test results and treatment outcomes for gram-positive mastitis pathogens following treatment with cephapirin sodium. *Journal of Dairy Science*, 92, 2589–97.

Barkema, H. W., Schukken, Y. H. and Zadoks, R. N. 2006. Invited review: The role of cow, pathogen, and treatment regimen in the therapeutic success of bovine Staphylococcus aureus mastitis. *Journal of Dairy Science*, 89, 1877–95.

Bengtsson, B., Unnerstad, H. E., Ekman, T., Artursson, K., Nilsson-Ost, M. and Waller, K. P. 2009. Antimicrobial susceptibility of udder pathogens from cases of acute clinical mastitis in dairy cows. *Veterinary Microbiology*, 136, 142–9.

Boerlin, P. and White, D. G. 2013. Antimicrobial Resistance and Its Epidemiology. In S. Giguere, J. F. Prescott and P. M. Dowling (eds), *Antimicrobial Therapy in Veterinary Medicine, 5th edition.* 5th ed., Ames, IA: Wiley Blackwell Publishing.

Bradley, A. J. and Green, M. J. 2009. Factors affecting cure when treating bovine clinical mastitis with cephalosporin-based intramammary preparations. *J. Dairy Sci.*, 92, 1941–53.

Bradley, A. J., Leach, K. A., Breen, J. E., Green, L. E. and Green, M. J. 2007. Survey of the incidence and aetiology of mastitis on dairy farms in England and Wales. *Vet Rec*, 160, 253–7.

Burvenich, C., Van Merris, V., Mehrzad, J., Diez-Fraile, A. and Duchateau, L. 2003. Severity of E. coli mastitis is mainly determined by cow factors. *Vet. Res.*, 34, 521–64.

Cicconi-Hogan, K. M., Belomestnykh, N., Gamroth, M., Ruegg, P. L., Tikofsky, L. and Schukken, Y. H. 2014. Short communication: Prevalence of methicillin resistance in coagulase-negative staphylococci and Staphylococcus aureus isolated from bulk milk on organic and conventional dairy farms in the United States. *J. Dairy Sci.*, 97, 2959–64.

Erskine, R. J., Cullor, J., Schaellibaum, M., Yancey, R. and Zecconi, A. 2004. Bovine mastitis pathogens and trends in resistance to antibacterial drugs. In N. M. Council (ed.), Annual Meeting National Mastitis Council, 2004 Charlotte, NC. Madison, WI: National Mastitis Council, 400–14.

Erskine, E. J., Walker, R. D., Bolin, C. A., Bartlett, P. C. and White, D. G. 2002. Trends in antibacterial susceptibility of mastitis pathogens during a seven-year period. *Journal of Dairy Science*, 85, 1111–18.

Garcia-Migura, L., Hendriksen, R. S., Fraile, L. and Aarestrup, F. M. 2014. Antimicrobial resistance of zoonotic and commensal bacteria in Europe: The missing link between consumption and resistance in veterinary medicine. *Veterinary Microbiology*, 170, 1–9.

Gindonis, V., Taponen, S., Myllyniemi, A. L., Pyorala, S., Nykasenoja, S., Salmenlinna, S., Lindholm, L. and Rantala, M. 2013. Occurrence and characterization of methicillin-resistant staphylococci from bovine mastitis milk samples in Finland. *Acta Vet Scand*, 55, 61.

Gonzalez Pereyra, V., Pol, M., Pastorino, F. and Herrero, A. 2015. Quantification of antimicrobial usage in dairy cows and preweaned calves in Argentina. *Prev. Vet. Med.*, 122, 273–9.

Haveri, M., Suominen, S., Rantala, L., Honkanen-Buzalski, T. and Pyorala, S. 2005. Comparison of phenotypic and genotypic detection of penicillin G resistance of Staphylococcus aureus isolated from bovine intramammary infection. *Vet. Microbiol.*, 106, 97–102.

Hektoen, L., Odegaard, S. A., Loken, T. and Larsen, S. 2004. Evaluation of stratification factors and score-scales in clinical trials of treatment of clinical mastitis in dairy cows. *J. Vet. Med. A Physiol. Pathol. Clin. Med.*, 51, 196–202.

Hoe, F. G. and Ruegg, P. L. 2005. Relationship between antimicrobial susceptibility of clinical mastitis pathogens and treatment outcome in cows. *J. Am. Vet. Med. Assoc.*, 227, 1461–8.

Intorre, L., Vanni, M., Meucci, V., Tognetti, R., Cerri, D., Turchi, B., Cammi, G., Arigonni, N. and Garabarino, C. 2013. Antimicrobial resistance of Staphylococcus aureus isolated from bovine milk in Italy from 2005 to 2011. *Large Animal Review,* 19, 287–91.

Jensen, V. F., Jacobsen, E. and Bager, F. 2004. Veterinary antimicrobial-usage statistics based on standardized measures of dosage. *Prev. Vet. Med.*, 64, 201–15.

Jones, P. J., Marier, E. A., Tranter, R. B., Wu, G., Watson, E. and Teale, C. J. 2015. Factors affecting dairy farmers' attitudes towards antimicrobial medicine usage in cattle in England and Wales. *Prev. Vet. Med.*, 121, 30–40.

Kuipers, A., Koops, W. J. and Wemmenhove, H. 2016. Antibiotic use in dairy herds in the Netherlands from 2005 to 2012. *J. Dairy Sci.*, 99, 1632–48.

Lago, A., Godden, S. M., Bey, R., Ruegg, P. L. and Leslie, K. 2011a. The selective treatment of clinical mastitis based on on-farm culture results: I. Effects on antibiotic use, milk withholding time, and short-term clinical and bacteriological outcomes. *J. Dairy Sci.*, 94, 4441–56.

Lago, A., Godden, S. M., Bey, R., Ruegg, P. L. and Leslie, K. 2011b. The selective treatment of clinical mastitis based on on-farm culture results: II. Effects on lactation performance, including clinical mastitis recurrence, somatic cell count, milk production, and cow survival. *J. Dairy Sci.*, 94, 4457–67.

Makovec, J. A. and Ruegg, P. L. 2003. Antimicrobial resistance of bacteria isolated from dairy cow milk samples submitted for bacterial culture: 8,905 samples (1994–2001). *J. Am. Vet. Med. Assoc.*, 222, 1582–9.

McDougall, S., Arthur, D. G., Bryan, M. A., Vermunt, J. J. and Weir, A. M. 2007. Clinical and bacteriological response to treatment of clinical mastitis with one of three intramammary antibiotics. *N Z Vet. J.*, 55, 161–70.

Mitchell, J. M., Griffiths, M. W., Mcewen, S. A., Mcnab, W. B. and Yee, A. J. 1998. Antimicrobial drug residues in milk and meat: Causes, concerns, prevalence, regulations, tests, and test performance. *J. Food Prot.*, 61, 742–56.

Moroni, P., Pisoni, G., Antonini, M., Villa, R., Boettcher, P. and Carli, S. 2006. Short communication: Antimicrobial drug susceptibility of Staphylococcus aureus from subclinical bovine mastitis in Italy. *J. Dairy Sci.*, 89, 2973–6.

Neu, H. C. 1992. An Update on Fluoroquinolones. *Current Opinion in Infectious Diseases*, 5, 755–63.

Oliveira, L., Hulland, C. and Ruegg, P. L. 2013. Characterization of clinical mastitis occurring in cows on 50 large dairy herds in Wisconsin. *J. Dairy Sci.*, 96, 7538–49.

Oliveira, L., Langoni, H., Hulland, C. and Ruegg, P. L. 2012. Minimum inhibitory concentrations of Staphylococcus aureus recovered from clinical and subclinical cases of bovine mastitis. *J.Dairy Sci.*, 95, 1913–20.

Oliveira, L. and Ruegg, P. L. 2014. Treatments of clinical mastitis occurring in cows on 51 large dairy herds in Wisconsin. *J. Dairy Sci.*, 97, 5426–36.

Oliver, S. P., Almeida, R. A., Gillespie, B. E., Headrick, S. J., Dowlen, H. H., Johnson, D. L., Lamar, K. C., Chester, S. T. and Moseley, W. M. 2004a. Extended ceftiofur therapy for treatment of experimentally-induced Streptococcus uberis mastitis in lactating dairy cattle. *J. Dairy Sci.*, 87, 3322–9.

Oliver, S. P., Gillespie, B. E., Headrick, S. J., Moorehead, H., Lunn, P., Dowlen, H. H., Johnson, D. L., Lamar, K. C., Chester, S. T. and Moseley, W. M. 2004b. Efficacy of extended ceftiofur intramammary therapy for treatment of subclinical mastitis in lactating dairy cows. *J. Dairy Sci.*, 87, 2393–400.

Oliver, S. P. and Murinda, S. E. 2012. Antimicrobial resistance of mastitis pathogens. *Veterinary Clinics of North America-Food Animal Practice*, 28, 165.

Petrovski, K. R., Grinberg, A., Williamson, N. B., Abdalla, M. E., Lopez-Villalobos, N., Parkinson, T. J., Tucker, I. G. and Rapnicki, P. 2015. Susceptibility to antimicrobials of mastitis-causing Staphylococcus aureus, Streptococcus uberis and Str. dysgalactiae from New Zealand and the USA as assessed by the disk diffusion test. *Aust. Vet. J.*, 93, 227–33.

Pinzon-sanchez, C., Cabrera, V. E. and Ruegg, P. L. 2011. Decision tree analysis of treatment strategies for mild and moderate cases of clinical mastitis occurring in early lactation. *J. Dairy Sci.*, 94, 1873–92.

Pinzon-sanchez, C. and Ruegg, P. L. 2011. Risk factors associated with short-term post-treatment outcomes of clinical mastitis. *J. Dairy Sci.*, 94, 3397–410.

Pol, M. and Ruegg, P. L. 2007a. Relationship between antimicrobial drug usage and antimicrobial susceptibility of gram-positive mastitis pathogens. *J. Dairy Sci.*, 90, 262–73.

Pol, M. and Ruegg, P. L. 2007b. Treatment practices and quantification of antimicrobial drug usage in conventional and organic dairy farms in Wisconsin. *J. Dairy Sci.*, 90, 249–61.

Prescott, J. F. 2006. History of antimicrobial usage in agriculture: an overview. In F. M. Aarestrup (ed.), *Antimicrobial Resistance in Bacteria of Animal Origin*. Washington, DC.: ASM Press.

Riekerink, R. G. M. O., Barkema, H. W., Kelton, D. F. and Scholl, D. T. 2008. Incidence rate of clinical mastitis on Canadian dairy farms. *J. Dairy Sci.*, 91, 1366–77.

Ruegg, P. L. 2011. Managing mastitis and producing quality milk. *Dairy Production Medicine*, 207–32.

Ruegg, P. L., Erskine, R. D. and Morin, D. 2014. Mammary Gland Health. In B. P. Smith (ed.), *Large Animal Veterinary INternal Medicine*, 5th ed., USA: Elsevier.

Ruegg, P. L., Oliveira, L., Jin, W. and Okwumabua, O. 2015a. Phenotypic antimicrobial susceptibility and occurrence of selected resistance genes in Gram-positive mastitis pathogens isolated from Wisconsin dairy cows. *J. Dairy Sci.*, 98, in press.

Ruegg, P. L., Oliveira, L., Jin, W. and Okwumabua, O. 2015b. Phenotypic antimicrobial susceptibility and occurrence of selected resistance genes in Gram-positive mastitis pathogens isolated from Wisconsin dairy cows. *J. Dairy Sci.*, 98, in press.

Ruegg, P. L. and Tabone, T. J. 2000. The relationship between antibiotic residue violations and somatic cell counts in Wisconsin dairy herds. *J. Dairy Sci.*, 83, 2805–9.

Saini, V., Mcclure, J. T., Leger, D., Dufour, S., Sheldon, A. G., Scholl, D. T. and Barkema, H. W. 2012a. Antimicrobial use on Canadian dairy farms. *J. Dairy Sci.*, 95, 1209–21.

Saini, V., Mcclure, J. T., Leger, D., Keefe, G. P., Scholl, D. T., Morck, D. W. and Barkema, H. W. 2012b. Antimicrobial resistance profiles of common mastitis pathogens on Canadian dairy farms. *J. Dairy Sci.*, 95, 4319–32.

Saini, V., Mcclure, J. T., Scholl, D. T., Devries, T. J. and Barkema, H. W. 2012c. Herd-level association between antimicrobial use and antimicrobial resistance in bovine mastitis Staphylococcus aureus isolates on Canadian dairy farms. *J. Dairy Sci.*, 95, 1921–9.

Saini, V., Mcclure, J. T., Scholl, D. T., Devries, T. J. and Barkema, H. W. 2013. Herd-level relationship between antimicrobial use and presence or absence of antimicrobial resistance in gram-negative bovine mastitis pathogens on Canadian dairy farms. *J. Dairy Sci.*, 96, 4965–76.

Sandberg, K. D. and Lapara, T. M. 2016. The fate of antibiotic resistance genes and class 1 integrons following the application of swine and dairy manure to soils. *FEMS Microbiol. Ecol.*, 92.

Schmidt, T., Kock, M. M. and Ehlers, M. M. 2015. Diversity and antimicrobial susceptibility profiling of staphylococci isolated from bovine mastitis cases and close human contacts. *J. Dairy Sci.*, 98, 6256–69.

Serraino, A., Giacometti, F., Marchetti, G., Zambrini, A. V., Fustini, M. and Rosmini, R. 2013. Survey on antimicrobial residues in raw milk and antimicrobial use in dairy farms in the Emilia-Romagna region, Italy. *Italian J. Animal Sci.*, 12.

Sol, J., Sampimon, O. C., Barkema, H. W. and Schukken, Y. H. 2000. Factors associated with cure after therapy of clinical mastitis caused by Staphylococcus aureus. *J. Dairy Sci.*, 83, 278–84.

Stevens, M., Piepers, S., Supre, K., Dewulf, J. and de vliegher, S. 2016. Quantification of antimicrobial consumption in adult cattle on dairy herds in Flanders, Belgium, and associations with udder health, milk quality, and production performance. *J. Dairy Sci.*, 99, 2118–30.

Suojala, L., Kaartinen, L. and Pyorala, S. 2013. Treatment for bovine Escherichia coli mastitis – an evidence-based approach. *J. Vet. Pharmacol. Ther.*, 36, 521–31.

Tenhagen, B. A., Vossenkuhl, B., Kaesbohrer, A., Alt, K., Kraushaar, B., Guerra, B., Schroeter, A. and Fetsch, A. 2014. Methicillin-resistant Staphylococcus aureus in cattle food chains – Prevalence, diversity, and antimicrobial resistance in Germany. *J. Animal Sci.*, 92, 2741–51.

Thomas, V., de jong, A., Moyaert, H., Simjee, S., El garch, F., Morrissey, I., Marion, H. and Valle, M. 2015. Antimicrobial susceptibility monitoring of mastitis pathogens isolated from acute cases of clinical mastitis in dairy cows across Europe: VetPath results. *Int. J. Antimicrob. Agents*, 46, 13–20.

USDA 2008. Dairy 2007, Part III: Reference of dairy cattle health and management practices in the United States, 2007. USDA-APHIS-VS, CEAH. Fort Collins CO.

Weese, J. S., Page, S. W. and Prescott, J. F. 2013. Antimicrobial stewardship in animals. In S. Giguere, J. F. prescott and P. M. Dowling (eds), *Antimicrobial Therapy*. Ames, IA: Wiley Blackwell.

Wageningen, 2013. http://www.wageningenur.nl/en/Research-Results/Projects-and-programmes/MARAN-Antibiotic-usage.htm.